Palgrave Macmillan Memory Studies

Series Editors: **Andrew Hoskins** and **John Sutton**

International Advisory Board: **Steven Brown**, University of Leicester, UK, **Mary Carruthers**, New York University, USA, **Paul Connerton**, University of Cambridge, UK, **Astrid Erll**, University of Wuppertal, Germany, **Robyn Fivush**, Emory University, USA, **Tilmann Habermas**, University of Frankfurt am Main, Germany, **Jeffrey Olick**, University of Virginia, USA, **Susannah Radstone**, University of East London, UK, and **Ann Rigney**, Utrecht University, the Netherlands

The nascent field of Memory Studies emerges from contemporary trends that include a shift from concern with historical knowledge of events to that of memory, from 'what we know' to 'how we remember it'; changes in generational memory; the rapid advance of technologies of memory; panic over declining powers of memory, which mirrors our fascination with the possibilities of memory enhancement; and the development of trauma narratives in reshaping the past.

These factors have contributed to an intensification of public discourses on our past over the last thirty years. Technological, political, interpersonal, social and cultural shifts affect what, how and why people and societies remember and forget. This groundbreaking series tackles questions such as: What is 'memory' under these conditions? What are its prospects, and also the prospects for its interdisciplinary and systematic study? What are the conceptual, theoretical and methodological tools for its investigation and illumination?

Silke Arnold-de Simine
MEDIATING MEMORY IN THE MUSEUM
Empathy, Trauma, Nostalgia

Rebecca Bramall
THE CULTURAL POLITICS OF AUSTERITY
Past and Present in Austere Times

Irit Dekel
MEDIATION AT THE HOLOCAUST MEMORIAL IN BERLIN

Anne Fuchs
AFTER THE DRESDEN BOMBING
Pathways of Memory 1945 to the Present

Irial Glynn and J. Olaf Kleist (*editors*)
HISTORY, MEMORY AND MIGRATION
Perceptions of the Past and the Politics of Incorporation

Andrea Hajek
NEGOTIATING MEMORIES OF PROTEST IN WESTERN EUROPE
The Case of Italy

Amy Holdsworth
TELEVISION, MEMORY AND NOSTALGIA

Jason James
PRESERVATION AND NATIONAL BELONGING IN EASTERN GERMANY
Heritage Fetishism and Redeeming Germanness

Sara Jones
THE MEDIA OF TESTOMONY
Remembering the East German Stasi in the Berlin Republic

Emily Keightley and Michael Pickering
THE MNEMONIC IMAGINATION
Remembering as Creative Practice

Amanda Lagerkvist
MEDIA AND MEMORY IN NEW SHANGHAI
Western Performances of Futures Past

Philip Lee and Pradip Ninan Thomas (*editors*)
PUBLIC MEMORY, PUBLIC MEDIA AND THE POLITICS OF JUSTICE

Erica Lehrer, Cynthia E. Milton and Monica Eileen Patterson (*editors*)
CURATING DIFFICULT KNOWLEDGE
Violent Pasts in Public Places

Oren Meyers, Eyal Zandberg and Motti Neiger
COMMUNICATING AWE
Media, Memory and Holocaust Commemoration

Anne Marie Monchamp
AUTOBIOGRAPHICAL MEMORY IN AN ABORIGINAL AUSTRALIAN COMMUNITY
Culture, Place and Narrative

Motti Neiger, Oren Meyers and Eyal Zandberg (*editors*)
ON MEDIA MEMORY
Collective Memory in a New Media Age

Katharina Niemeyer (*editor*)
MEDIA AND NOSTALGIA
Yearning for the Past, Present and Future

Margarita Saona
MEMORY MATTERS IN TRANSITIONAL PERU

Anna Saunders and Debbie Pinfold (*editors*)
REMEMBERING AND RETHINKING THE GDR
Multiple Perspectives and Plural Authenticities

V. Seidler
REMEMBERING DIANA
Cultural Memory and the Reinvention of Authority

Bryoni Trezise
PERFORMING FEELING IN CULTURES OF MEMORY

Evelyn B. Tribble and Nicholas Keene
COGNITIVE ECOLOGIES AND THE HISTORY OF REMEMBERING
Religion, Education and Memory in Early Modern England

Barbie Zelizer and Keren Tenenboim-Weinblatt (*editors*)
JOURNALISM AND MEMORY

Palgrave Macmillan Memory Studies
Series Standing Order ISBN 978-0-230-23851-0 (hardback)
978-0-230-23852-7 (paperback)
(*outside North America only*)

You can receive future titles in this series as they are published by placing a standing order. Please contact your bookseller or, in case of difficulty, write to us at the address below with your name and address, the title of the series and the ISBN quoted above.

Customer Services Department, Macmillan Distribution Ltd, Houndmills, Basingstoke, Hampshire RG21 6XS, England

Performing Feeling in Cultures of Memory

Bryoni Trezise
School of the Arts and Media, University of New South Wales, Australia

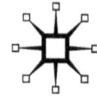

© Bryoni Trezise 2014

All rights reserved. No reproduction, copy or transmission of this publication may be made without written permission.

No portion of this publication may be reproduced, copied or transmitted save with written permission or in accordance with the provisions of the Copyright, Designs and Patents Act 1988, or under the terms of any licence permitting limited copying issued by the Copyright Licensing Agency, Saffron House, 6–10 Kirby Street, London EC1N 8TS.

Any person who does any unauthorized act in relation to this publication may be liable to criminal prosecution and civil claims for damages.

The author has asserted her right to be identified as the author of this work in accordance with the Copyright, Designs and Patents Act 1988.

First published 2014 by
PALGRAVE MACMILLAN

Palgrave Macmillan in the UK is an imprint of Macmillan Publishers Limited, registered in England, company number 785998, of Houndmills, Basingstoke, Hampshire RG21 6XS.

Palgrave Macmillan in the US is a division of St Martin's Press LLC, 175 Fifth Avenue, New York, NY 10010.

Palgrave Macmillan is the global academic imprint of the above companies and has companies and representatives throughout the world.

Palgrave® and Macmillan® are registered trademarks in the United States, the United Kingdom, Europe and other countries.

ISBN 978–1–137–33621–7

This book is printed on paper suitable for recycling and made from fully managed and sustained forest sources. Logging, pulping and manufacturing processes are expected to conform to the environmental regulations of the country of origin.

A catalogue record for this book is available from the British Library.

A catalog record for this book is available from the Library of Congress.

Transferred to Digital Printing in 2015

To Mum and Dad, with love and thanks

Contents

List of Figures viii

Acknowledgements ix

Introduction: Feeling the Return of Memory 1

1 Sensing the Holocaust Affect: Memorials in Repeat, Revision and Return 28

2 Becoming Other-wise: Remembering Intercorporeal Indigeneity *Down Under* 56

3 Feeling Remediated: The Emotional Afterlife of Psychic Trauma TV 79

4 Affecting Indifference: Traumatic A-materiality in Second Life 107

5 Affect's Spill: Theatrical 'Sensationship' in Cultures of Memory 136

Notes 158

Bibliography 180

Index 192

Figures

I.1	Artur Żmijewski *80064*, 2004	1
1.1	Jane Korman, 'I Will Survive', *Dancing Auschwitz* (2010)	28
1.2	Nandor Glid, *International Monument* (1968) at Dachau Memorial Concentration Camp	34
1.3	*Missing House*, Christian Boltanski (1990)	43
1.4	Menashe Kadishman, Memory Void Installation *Shalechet* (view to the passageway), (2001)	47
1.5	*Memorial to the Murdered Jews of Europe*, Peter Eisenman (2005)	50
1.6	Jane Korman, 'Art Must Go On', *Dancing Auschwitz* (2010)	53
4.1	Digitised photograph: Germans pass by the broken shop window of a Jewish-owned business, screenshot Kristallnacht in Second Life	121
4.2	Kristallnacht streetscape, screenshot Kristallnacht in Second Life	123
4.3	Kristallnacht Witness Rotunda, screenshot Kristallnacht in Second Life	125
4.4	Aryanised shop, screenshot Kristallnacht in Second Life	130
4.5	Hidden Room, screenshot Kristallnacht in Second Life	133
5.1	*Tragedia Endogonidia BR.#04. Bruxelles*, photograph by Luca Del Pia	136
5.2	*Tragedia Endogonidia BR.#04. Bruxelles*, photograph by Luca Del Pia	148

Acknowledgements

There are many histories of my own that sit between the words that follow in these pages, reminding me just how long this work has been with me. Its stretch recalls the early days of writing my PhD, and for that part of its evolution I am indebted to my supervisor Edward Scheer, as well as to its assessors, including the supportive Helena Grehan, who gave important critiques that have inevitably shaped its afterlife. At its other bookend, I see my pencilled underlining in various books as traces of the arrival of my daughter Bela: her sleeping in the car, me trying to grab a thought in the front seat. Feeling her burst of energy in the world has, of course, made memory into its own kind of history, the force of her creativity pushing me well into the future.

Before Bela, there was a tenuous journey supported by colleagues who I am privileged to call my friends: Clare Grant, Meg Mumford and Erin Brannigan. I owe thanks to each for the wisdom they have shown me and for their acuity and wit, and of course, presence to the world of performance. I would also like to thank George Kouvaros and Ursula Rao for generously helping to shape the framework of the project in its early stages.

Sections of the book have been published in other places and have been reworked here specifically for how they draw out new theoretical terrain. Small sections from Chapter 1 first appeared in chapters of *Performance, Embodiment and Cultural Memory* (Cambridge Scholars Press, 2009) and *Performing Feeling in Cultures of Memory* (Museum Tusculanum Press, 2013). Sections from chapters 3, 4 and 5 first appeared in the journals *Performance Research*, *Memory Studies* and *Theatre Research International*, respectively. I am grateful to the publishers for permission to include these sections and to the anonymous reviewers who enhanced earlier drafts. Some of this work has also been presented at conferences held by Performance Studies International and the Australasian Drama Studies Association. I am appreciative of the generous feedback respondents have given me along the way. The School of the Arts and Media and the Faculty of Arts and Social Sciences at the University of New South Wales have supported this work by affording me time as well as funds to refine my thinking, and I am acutely aware of the difference this has made.

Untold thanks go to a dear colleague and friend, Caroline Wake, whose work has inspired my own, and who helps to keep my feet on the ground and my chin up with our great many *chats*. Thanks also to her sister, Neph Wake, for lending an acute eye to the book's cover design. I would also like to thank my old and dearest friends Evelyn Harvey and Kate Solomon for being expert witnesses to the chaos in my life for what feels like too long. My parents and brother have watched carefully as I fashioned myself into an academic, swallowing time and other pathways in the process. I'm indebted to them for fostering in me the courage to think and create; I have here what can only seem a small offering in return. And to Pablo, the ever-mindful, ever-present collaborator in my life who expands my world and makes it full.

Introduction: Feeling the Return of Memory

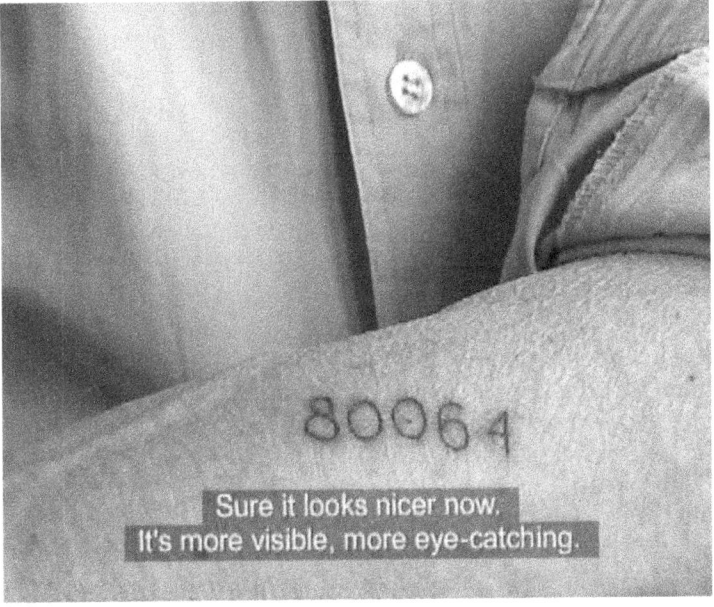

Figure I.1 Artur Żmijewski *80064*, 2004
Single channel video, projection or monitor, 11 min., colour, sound. Ed. of 3 + 1 AP + 1 EP. Videostill.
Source: Courtesy of the artist, Galerie Peter Kilchmann, Zurich, and Foksal Gallery Foundation, Warsaw.

Memory's affect

When Polish artist Artur Żmijewski asked 92-year-old Józef Tarnawa, a former prisoner of the Auschwitz-Birkenau concentration camp, to 'restore' the camp number tattooed on his forearm, Żmijewski's film of the event, *80064*, evoked an unsettling complicity in spectators. Żmijewski has since explained that he aimed for a 'film experiment with memory', that 'under the effect of the tattooing the "doors of memory" would open, that there would be an eruption of remembrance of that time, a stream of images or words describing the painful past'.[1] Tarnawa, who initially agreed to the proposed 'redoing' of the wound, appears uncomfortably reluctant on screen. The reinscription is 'not necessary', he repeats, for 'it won't be the same number, it will be restored'. Żmijewski persists and, by the close of the film, Tarnawa seems unconvincingly pleased with the updated version: 'It looks nicer now, it's more visible, more eye-catching...I have renovated it like some piece of furniture.'[2] As Claire Bishop has remarked, Żmijewski's decided troubling of the traumatic wound addresses 'ethics as an explicit theme': it makes a point of our role in watching the wound become a site of repetition.[3] As we watch, we are bystanders to Żmijewski's desire to incite a performance of memory, and while that performance misfires, we nonetheless become part of his anticipation that the 'Real' will erupt and touch us with its feeling.

In the same way that Elaine Scarry has made us aware of pain, Żmijewski seems to want to make clear that feeling happens in the present, even if it is recalling a past.[4] Perhaps his film signals most strongly that the return of memory as an eruption of feeling invites us to become players in the spectacle of traumatic acting out. His camera, focussing closely on the skin as it is inscribed number by number, and then cutting to close-ups, which pause on Tarnawa as if the tattoo would draw something else out of the depths of his body, stages and incites this drama of vision. The catch is that Tarnawa thwarts the frame; he refuses to perform memory. He is instead concerned with the ongoing performance of the wound itself, with how the wrinkled marker of history that evidences his having been there will continue to illuminate itself as witness once restored. Żmijewski's logic that the spectacle of trauma can be triggered by an act of sensory repetition is not his alone. His film makes a point of the role of the body in transmitting memory and of its framing as the property of a public who expect it to communicate in exceptionally affective ways. With our gaze focused on Tarnawa and our bodies at the ready to receive his embodied insight, our looking backfires as he instead experiences his own role in relation to the camera. Tarnawa resists performing as a site of our affective property. This is a political move that unsettles the alignment presumed between spectator and subject. While we don't feel his memories, we feel our own short-circuited feelings in other, perhaps more complex ways.

In this book I draw on art-makers such as Żmijewski, and a range of other live, televisual, digital and theatrical cultural practices, to investigate the role of feeling in socially mediating collective histories of trauma. More precisely, I understand that the transaction between historical subject and contemporary spectator is one that is created through practices that attempt to arrest, repeat and rehearse traumatic affect as a means of creating cultural memory, practices which inevitably produce the subject formations of wounded and witnessing selves. Feelings, as both the conduit of experience and the way that we experience our experience, have effects for the future and on behalf of the past. They bind, separate and differentiate, but also connect and contaminate. Feelings happen in bodies and to bodies, but they also happen as a process of self-making. How I feel determines my orientation to the world and my orientation to other feeling bodies. My feelings, although they feel as if they are solely mine, are my body's process of interactive self-perception. As Sara Ahmed puts it, emotions – as one form of feeling – don't merely happen, they are 'done', 'bound up with the sticky relation between signs and bodies: emotions...materialise the surfaces and boundaries that are lived as worlds'.[5] My body knows itself and feels itself through the objects, places, smells and other bodies that it encounters. Via writers such as Ahmed, I ask how cultures of memory engineer the ways that spectators feel for and about difficult pasts. In this sense, I understand feelings as memory practices that have, and are, particular cultural effects.

In the chapters that follow, I argue that the lived relations produced by the cultural narration of traumatic memory can be most fundamentally understood through a focus on how spectators sense and become sentimental about the past. Correlatively, I suggest that those lived relations, which are always felt, embodied relations, engage assumptions around the 'feel-ability' of certain bodies over others. In short, they exploit the affective charge that a wounded or traumatised subject is culturally given to possess. Feeling, as an experience in the present that mediates the past, then comes to more implicitly mediate all kinds of social and cultural difference.

Feel-ability accounts for the assumed capacity of one body's history to be felt by another, as well as for the assumed moral certitude in doing so. Feel-ability can even point to those of us who might, on occasion, feel good about feeling bad for someone else. I therefore ask how and why feeling is positioned at the centre of our expressions of cultural memory. Sensuous complicities come to model the means and consequences by which we participate in variously tactile, sensory and emotive forms of recollection.

While memory has been at the forefront of critical debate in the humanities since the 1980s, the affective turn, as a more recent means to think about how bodies work in, and as, culture, has not yet been used to concretely illuminate it. In this book I chart a series of memory sites that mobilise various histories of loss through the experiences and re-experiences of affect. These sites range from established, even canonical museums and memorials, to those practices enabled by the virtual terrain of Second Life, to the more implicit affective exchanges that are produced by popular 'trauma TV' programmes or the aesthetics of radical, contemporary theatre practice. Each memory site focuses on the spectator's encounter with memory as an aesthetic phenomenon, and in this way, each site foregrounds the affective construction of 'pastness' as central to the success of its political effects. The sites together build on a vast field already committed to examining contemporary cultures of memory and, more specifically, on a field which has found rich synergies between languages of performance and the body's seminal role in carrying, transforming and producing memory effects. In this book I bring these discussions into conversation with a more recent focus on feelings as cultural actors. In doing so, I map conjunctions between the individual and institutional acts that comprise memory culture and the responses – considered here as practices – of affect and emotion that tend to accompany them.[6] Together, these critical frames amass a cultural politics and a critique that is built around feeling the memories of others.

In using the word 'feeling', I draw on Sara Ahmed's account of the wedding of physiology with emotion. Ahmed proposes the notion of the 'impression' to sidestep having to distinguish between the cognitive and sensory aspects of feeling. 'Impression', which denotes both a skinly mark and an affective charge or cognitive response 'avoid[s] making analytical distinctions between bodily sensation, emotion and thought as if they could be "experienced" as distinct realms of human "experience" '.⁷ While understanding that there are specific traditions and often contradicting inflections that inform current characterisations of embodiment and affect, and emotion and feeling, in simple terms I aim to foreground how feeling – as a decidedly corporeal 'impression' that is at once psychic, somatic and sensorial – operates as a relational process of subject formation at the same time as it works to hide this formative process from corporeal view. That is, I aim to foreground how feelings promise to make remembering and remembered subjects culturally distinct at the same time as they contaminate the very bodies they hope to individuate. Feelings possess us such that we feel ontologically unique, separate and self-bound in the very moment that we might just be radically, affectively open.

To this end, I argue that contemporary memory cultures are overinvested in an 'experiential turn', which operates to assume the virtue, or importance, of some feeling the memories of others. That is, feeling, as it is mobilised as a project of memory cultures, produces ideals of subjective certainty – of wounded and witness – when its processes often undo those parameters of being. At the same time, a process of feeling the memories of others often fails to acknowledge the constituting effects it produces. I call these forms of complicit engagement with the sensory past 'unthinking feeling' to prompt what might give way when processes of feeling begin to become rethought. In turn, I call the kind of corporeal rethinking that can make visible (or visceral) memory culture's sensory coercions 'affect's spill', and suggest that the spill of affect describes those rare aesthetic moments, as in Żmijewski's film, which use feeling to undo itself.

Affect's spill describes that sensory instant when affect punctures the political dimensions of affectivity. The broader question here is in why memory cultures privilege feeling as a means of generating an ethics of recognition, particularly as it reinforces what Ahmed terms a 'politics of pain'. As Ahmed explains, 'pain is involved in the production of *uneven* effects in ... that pain does not produce a homogeneous group of bodies who are together in their pain'.[8] Her focus on effects – the discursive manipulations of pain – rather than pain's interiority, shifts Scarry's reading of pain as essentially unsharable.[9] Ahmed makes it clear that however lonely pain might be, its 'labour' and ' "language" ... work in specific and determined ways to affect differences between bodies'.[10] In Vicki Kirby's terms, this kind of difference is always and already a ' "becoming entity": it is not a name for the gap of supposedly dead space and time between pregiven entities'.[11]

I hence suggest that a politics of producing ideals of otherness comes about as a process of feeling for and about the past. These ideals are intrinsic to the commodifying aspects of memory cultures and to the collective memory practices that those cultures produce. I draw together insights brought by Ahmed with those of Judith Butler, whose observations on loss explain that it is not only responsive to, but constitutive of, 'social, political and aesthetic relations'.[12] Butler helps us to see that while loss is a fundamental part of the human condition, it is also a social process that produces feelings of, and about, others. In this book I use thinkers like Butler and Ahmed to invert some of the dominant paradigms operating in the disciplines of memory and performance studies by understanding how practices of feeling produce the social categories they often purport to represent. In this sense, the book aims to unravel what both disciplines tend to assume about the practitioning of cultural memories and their respective investments in it. At its close, the book poses an antidote to 'unthinking feeling' by locating cultural memory practices that enable processes of feeling to be consciously re-felt. These processes can be understood as enabling the 'feeling' of feelings to occur: moments when affect spills out of, or into, itself. In the concluding chapter, I call these forms of culture 'meta-affective' and argue that the re-perceptions of loss that such aesthetic strategies produce can vitally enliven our bodies and minds to other ways of doing, thinking, feeling and being.

Repetition impression

While the tattoo reinscription in Żmijewski's film aimed to reopen memory's affective plenitude for victim and spectator, a more recent corporeal echo of his artistic experiment can be found in the descendants of Holocaust survivors themselves, who are now tattooing their arms with original prisoner identification numbers, in homage to their relatives. A *New York Times* article elaborates:

> When Eli Sagir showed her grandfather, Yosef Diamant, the new tattoo on her left forearm, he bent his head to kiss it. Mr. Diamant

had the same tattoo, the number 157622, permanently inked on his own arm by the Nazis at Auschwitz. Nearly 70 years later, Ms. Sagir got hers at a hip tattoo parlor downtown after a high school trip to Poland. The next week, her mother and brother also had the six digits inscribed onto their forearms.... 'I decided to do it to remind my generation: I want to tell them my grandfather's story and the Holocaust story [said Ms. Sagir, 21].'[13]

As Laurie Beth Clark has noted, the logic of 'never forget' can be a powerful cultural coercive that underpins a great many memory trends, institutions and acts.[14] How we remember, however, is often the less considered aspect of this central performative missive. Contrary to Tarnawa's renovated tattoo, these cross-generational tattoos might be seen to detach the sign of trauma from its bodily context and to reframe its signifying function. It could be suggested that in displacing and replacing the corporeal home of the original wound, the cross-generational tattoo aims to preserve and release the affective dimensions of the past, marking what the tattoo inheritor can never quite feel or know as yet another kind of feeling.

In Paul Connerton's study of the embodied practices of collective memory, tattoos exist 'at one and the same time *on* the surface of the skin and *under* the surface of the skin'. They constitute a 'physical boundary', which makes possible 'the capacity to constitute... the difference between cultural inclusion and exclusion'.[15] They demonstrate how memory '*takes place* on the body's surface and in its tissues' as a kind of 'semiosis [containing]... its own articulateness, history and purposes'.[16] Connerton highlights the role of the skin as central to developing what tattoo artist Vyvyn Lazonga calls a 'person's awareness of *memory*:... tattoos become points of reference that reinforce the self and its history'.[17] As a locus of touch, their inside-outness is a visual display of a process of feeling difference. At the same time, their *outside-inness* is at once visual and embodied, a sign through which the body feels itself, and is felt by others, as either 'outside' or 'in'. For Connerton, tattoos mark acts of honour or shame. It could be said that Holocaust tattoos mark the conjunction of both by performing the body's mastery over catastrophic time. The mark of shame has been transformed by time into a mark of honour and it is the body that has championed this feat of survival: the body has allowed the tattoo to survive *into* time. The Holocaust tattoo is evidence of the stretch of time and of the labour involved in surviving into the now.

While we understand that cross-generational Holocaust tattoos aim to bring history into the body as memory, and in so doing, to transfer the burden of the wound, we also might conjecture that as an act of transfer, it adulterates what we understand of memory. The tattoos might be seen to stage a skinly 'inter(in)animation', to use Rebecca Schneider's words, where '*then* and *now* punctuate each other'.[18] In her assemblage of a permanent inky leap from the 'here, now' of the hip tattoo parlour to the 'back then' of her grandfather's past, Eli Sagir stages, as does Żmijewski, a yearning to recuperate the singular knowledge contained in memory's affective social force. Like Żmijewski, however, she fails. As the sign of a sign, the cross-generational tattoo troubles the outside/in dichotomy proposed by Connerton because what it wears on its surface belongs to the impressive force experienced by another person's skin. What is internal to an imitation Holocaust tattoo is the shape of the original wound, but not its differentiating function. It stages *outside-outness*, and it uses that outness to create categories of who belongs 'in' and 'out'. As one literalisation of the desire to transfer an affective charge of the past through an experience in the present, the tattoo risks reproducing sign without substance; visible marker without tangible history. What it does do, however, is corporeally revision a new formation, or assemblage, of memory as a process and practice of feeling.

As an almost uncanny premonition of this now popular cross-generational practice, Żmijewski's film focalises the Holocaust tattoo as a cultural sign that impresses histories of social trauma upon us, unsettling us by making us responsible for the viewing lens it constructs. It identifies an uneasy enmeshment between historical contexts of Holocaust history and the broader visual cultures that now seek to bear witness to it, squeezing us between the affective hold of Tarnawa's retelling, and feeling the force of that hold as it produces us and as we have produced it. It is precisely this acute rupture between feeling sentiment out of sensation, and feeling the framework of that sentimental sensorium, that the film engineers as a space of critical awakening. In fact, and as I will aim to suggest throughout this book, it isolates the cultural space and significance of the political dimensions of memory's affectivity and, importantly, redirects this as a felt experience incurred by the film itself. This felt experience is what sears spectators of the work, who, tautologically, feel affect's affective dimensions. These are those fault lines that hide the ways in which feeling, as an intersubjective process, is used to produce sensations of subjective certainty. If *80064* speaks to those cultural practices that attempt to make history experientially close to some bodies on behalf of others, then it sharply reveals how feeling can produce subject-ideals, while its processes also undo those parameters of being.

As I will discuss in this book, the contexts with which Żmijewski's film is concerned participate in a form of culture that produces what Judith Butler calls the 'loss of loss'.[19] While the imitation tattoo remixes time as well as bodies and histories, it also makes claims about those whose job it is to do the work of memory, and those whose job it is to surrender the traces of their histories for that purpose. Butler explains that loss frames the human condition. For Butler, the grief that follows loss is a demonstration of 'the thrall in which our relations with others hold us, in ways that we cannot always recount or explain...in ways that challenge the very notion of ourselves as autonomous and in control'.[20] As Butler asks, 'Who "am" I without you?'. While it is the pull of time that causes us to wonder how we might make loss remain, Butler's suggestion that loss is enigmatic – that 'something is lost within the recesses of loss'[21] – invites us to think more deeply about how the personal condition of loss operates alongside its social production. Here, the losses recorded by the histories of exile, genocide, slavery and colonisation that Butler evokes are named against the observation that in recollecting loss, something that is central to it is also lost. In this case, the re-tattoo becomes a prompt to consider the nature of derivative feelings – what feelings about another's feelings *feel like*, and further, what other cultural effects are created by this specific kind of sensorial precarity. As I aim to show, with the loss of loss emerges the new cultural currency, and phenomenon, of what I call the memory affect.

The memory affect

The memory affect can be understood as a quality of reminiscence that invokes a feeling of embodied recall that does not locate a prior experience at its basis. It is the sign of a lived event that, dislocated from its original embodied source, is structured as something that *can* or *ought* to be 're-felt'. It can be partially understood through Marianne Hirsch's account of postmemory, which considers practices that emerge from the intergenerational transmission of trauma. It also connects to Lisa Saltzman's observations of post-Holocaust visual art forms that displace memory as a material index.[22] In stretching between what Hirsch outlines as the intensely familial experience of inheriting and often disowning the legacy of difficult pasts, and Saltzman's broader observations of the postmodern circularity of affect as a self-driven circuit of consumption and catharsis, the memory affect offers a means to think through the specific mode of corporeality that uses the presence of feeling to mediate and remediate *other* feelings (those said to be of, and in, the past). In other words, the memory affect challenges arguments which hold that 'feeling' the memories of others offers an automatically progressive means by which to recognise their traumatic pasts, and also points to the staging of this very moment of *re*-feeling as a distinct kind of experiential paradigm that operates on new terms. As the naming of a new social performative, it thereby conveys the means and effects by which spectator bodies become responsible and responsive to histories of trauma, and reveals that feelings can be designed to *feel as if they are the property of the body*, even when they are prompted by forces that feel 'supra-' or 'extra-' bodily.

Current theorisations of affect, embodiment, feeling and emotion place varying emphases on how (or whether) affect may be differentiated from feeling and emotion, and how (or whether) affect informs a practice and politics of sociality and subjectivity. Michael Hardt explains that two antecedents underpin affect's current critical divergences; the first offers a 'focus on the body, which has been most extensively advanced in feminist theory', and the second offers an 'exploration of emotions, conducted predominantly by queer theory'.[23] In Hardt's reading, both point to the importance of affect as offering a means to 'enter the realm of causality', such that we can use it to figure 'our power to affect the world around us and our power to be affected by it'.[24] Hardt's reading, along with that given by Patricia Ticineto Clough, points to affect as a 'substrate of potential bodily responses', as well as to the intrinsic sociality of that inner field. It is in this way for Clough 'not "presocial," as [Brian]

Massumi argues',[25] but entirely social in the sense that it can open out 'the economic circulation of preindividual bodily capacities or affects in the domain of biopolitical control'.[26] Affect in this sense cannot be thought of as anything other than the production of a sociality, one in which bodies are the agents and producers of affect's *affective* and *affecting* dimensions. This is, importantly, a form of sociality that affect, as an experiential determinant, comes to corporeally deny.

Common to affect's theoretical lineage here is its designation as a state of pre-individual intersubjectivity. In Brian Massumi's words, '[a]ffects are *virtual synesthetic perspectives* anchored in... the actually existing, particular things that embody them'.[27] It thereby designates '*a form of relation* as a rhythm, a fold, a timing, a habit, a contour, or a shape [which] comes to mark the passages of intensities... in body-to-body/world-body mutual imbrication'.[28] Using the analogy of a football game, affect is 'the mode in which potential is present in the perceiving body' such that it is 'a channeling of field-potential into local action'.[29] In this, bodies continuously unfold: when in motion the body 'coincides with its own transition: its own variation'.[30] Via Massumi, affect is neither sensation nor emotion but the prelude to both as they occur as lived expressions that participate in affect's movement. Those intensities, via Massumi (via Deleuze) are abstract forces that become '*materially* determining' (that is, *felt* or *acted upon*).[31] In this model, intensities pass between bodies to form boundaries between body and body, and body and world. That is, affects, as informers of moods, emotions and sentiments, are the precursors to the formation of the sense of self as it senses itself unfolding (with, or through) those sensing, or sensed, capacities. In Eric Shouse's terms, it is the very fact that affect is 'unformed and unstructured... that makes it transmittable in ways that feelings and emotions are not, and it is because affect is transmittable that it is potentially such a powerful social force'.[32]

Eve Kosofsky Sedgwick's alternately psychobiological model (via Silvan Tomkins) maps how drives of the human motivational system initiate affect's cultural transmission. As Anna Gibbs explains, in this model, intensities work from inside out, such that 'affects... are innate activators of themselves (fear makes the hair stand on end which produces more fear)'.[33] They also share themselves such that the 'biological capacities for affective response, mimetic communication and cross-modalization are co-opted in and by the cultural world'.[34] This propensity of affect as a primarily intersubjective force of contagion is most compellingly charted in Teresa Brennan's study of psychodynamics, which brings previously differentiated domains of psyche and soma under a framework that understands feelings as at once psychosomatic and biosocial. As one facet of her wide-ranging study, Brennan observes hormonal and olfactory processes which involve the entrainment of one body within another body, to explain how we experience the uncanny sense that ' "atmosphere" or the environment literally gets into the individual'.[35] For Brennan, affect becomes a staging or presencing *in* the body (*of the other as the self in the body*) that is fundamentally material, and whose materialism is fundamentally intersubjective. That is, affects are the differentiating product of transmission between undifferentiated bodies: 'the emotions or affects of one person, and the enhancing or depressing energies these affects entail, can enter into another'. Here, the transmission of affect importantly 'undermines the dichotomy between the individual and the environment and the related opposition between the biological and the social'.[36]

For my purposes, these models speak to the contexts of a memory culture that attempts to impress the experiences of the past on some, on behalf of others. They further point out that the way in which affect is aesthetically deployed, and the process by which it aesthetically works, are incongruous. They do so by highlighting how affect operates as a pre-social corporeal register, which, for Brennan, Clough, and to some extent Shouse, comes to inform our understandings of the field of the social. They make clear that differences between bodies are not pre-inscribed parameters but emerge from the *betweenness* of bodies, and that betweenness emerges from how bodies *sense* or *experience* themselves as unitary and contained through what are fundamentally relational practices. Insights on affect as the means by which bodies presence themselves in relation to others become deeply important to understanding those cultural memory practices that engage spectators on behalf of other (past) bodies. Here, as Brennan explains, we understand that the Cartesian self is produced as an experience through processes that paradoxically undermine it, where it is

> structured in such a way as to give a person the sense that their affects and feeling are their own, and that they are energetically

and emotionally contained in the most literal sense. In other words, people experience themselves as containing their own emotions.[37]

In this reading, a practice that involves re-experiencing the memories of others presumes that the affective contagion of that other pre-exists the encounter that might instead produce it.

Memory cultures, losing loss

In my argument, the memory affect is a means for describing the corporeal process of sensing a past (and the intensities of emotion, sensation or sentiment that might arrive through this process), as well as the cultural process of circulating that field of sensation as a kind of ongoing potentiality. The memory affect as a concept thus aims to draw into focus memory's affective dimensions as they operate as socially causal actors that circulate 'preindividual bodily capacities... in the domain of biopolitical control'.[38] In this way, the memory affect is located specifically in contemporary, Western memory cultures, which, through such processes of circulation, produce specific forms of cultural memory that are highly affective and affecting.

The phrase 'memory culture' itself foregrounds those practices which justify the social function and import of recollection. That is, memory culture points to the social, institutional, textual and aesthetic *doings* that produce varying forms of cultural memory as a truth-effect. Wulf Kansteiner and Richard Crownshaw distinguish between the terms 'collective memory' and 'cultural memory', and both have noted their debt to the writings of Jan Assman, who studied under sociologist of memory Maurice Halbwachs.[39] For Crownshaw, where 'collective' tends to emphasise communal memory processes, the term 'cultural' emphasises how memory becomes communicated, the 'means of its transmission'.[40] For Kansteiner, the term 'cultural' underscores the 'materiality of memory', and that memory is itself creative of what he terms 'objectified culture': 'the texts, rites, images, buildings and monuments which are designed to recall fateful events in the history of the collective'.[41] The phrase 'memory culture' positions memory as an adjective rather than a noun, as the descriptor for culture rather than as one cultural aspect out of many. It also denotes that the processes, acts and behaviours that constitute contemporary life are those that are done *about* and *through* memory. In this book, I understand that memory culture is the result, practitioning of, and discourse around which cultural – or collective – memory operates.[42]

Descriptions of memory culture observe the West in the grip of a memory epidemic, one which is witness to crises of authenticity and generative of amnesia. These arguments largely extend from observations of postmodernity as the primary free market mode. Frederic Jameson's model stresses the driving force of capital in all forms of cultural production: memory culture is a propellant of this economy and created by it. Andreas Huyssen finds these traits in the habitual building of 'museums and memorials as if haunted by the fear of some imminent traumatic loss', repetitions which persist against the inevitable 'forgetting of memory itself: nothing to remember, nothing to forget'.[43] Pierre Nora likewise famously lamented the loss of remembering brought about by postmodern times: 'We speak so much of memory', he wrote, 'because there is so little of it left.'[44] While Huyssen finds amnesia in 'the historicizing restoration of old urban centers, whole museum villages and landscapes, the boom of flea markets, retro fashions, and nostalgia waves',[45] for Nora, these restrict contexts for 'true' memory and externalise what was once organic. 'Real environments' of memory (*milieux*) have become 'sites' of memory (*lieux*). Housed in 'museums, archives, cemeteries, festivals, anniversaries, treaties, depositions, monuments, sanctuaries, fraternal orders', they are 'moments of history torn away from the movement of history... no longer quite life, not yet death'.[46]

In these ideas, memory culture defines not merely an archival obsession, but a social habit of interacting with remains. It produces a field of signs that are not of a particular past, but that fetishise 'pastness'. What is allegedly lost is the body's organic capacity to remember; what are instead produced are a series of material sites that in different ways restage the body as a mechanism given over to remembering. In this sense, while Nora depicts the loss of organic memory, and while Huyssen depicts an archival culture destined for repetition, what remains at the scene of memory is the spectator body in its role as a rememberer. This is a body that is put to re-feel the so-called memories of others, and within this, the so-called *loss of the faculty of recollection*, in new ways. Wulf Kansteiner explains that we are yet to fully consider how memory 'consumers use, ignore, or transform... artifacts according to their own interests'.[47]

These less-acknowledged forms of experience can involve, as Rebecca Schneider notes, 'a body sitting at a table in an archive, bent over an "original" manuscript or peering at a screen, interacting with history'.[48] Richard Crownshaw similarly agrees that certain kinds of re-embodiment do take place in such processes, but stipulates that these are 'not the same kind of embodied experience found in communicative and episodic memory and its transmission'.[49] In these views, memory culture becomes responsible for the loss of embodied memory and productive of the public stage that aims to restore it.

Since the growth of the field of cultural memory and its academic study, considerations of the forms of ethical engagement with the traumatic histories of others have contemplated not only the loss of embodied memory, but also, alternately, the possibility of its transmission between bodies. As one facet of this conversation, Marianne Hirsch's notion of postmemory describes the space of transfer that delineates the intergenerational working-through of inherited pasts. It is specifically familial, accounting for

> the relationship of children of survivors of cultural or collective trauma to the experiences of their parents, experiences that they 'remember' only as the narratives and images with which they grew up, but that are so powerful, so monumental, as to constitute memories in their own right.[50]

In its echo of Cathy Caruth's reading of trauma as a belated experience, postmemory signals the way that memory's wandering referent supersedes the material bodies in which it originated. It offers an account of how communities who are born into the aftermath of trauma remix that history to address its place in their own lives. Hirsch explains that postmemory can play with fact and fiction, holding the imperative 'to (re)build and to mourn'.[51] Its basis is thereby 'in displacement... vicariousness and belatedness',[52] qualities which transcend memory's basis in lived experience. Postmemory troubles memory as it has been understood and offers a proposition for how relationships with those who have suffered trauma can be negotiated and addressed.

The promise of Hirsch's account is that opening the ownership of traumatic cultural spaces can enable the shared working-through of inherited traumas and historically untouchable zones. Writers such as Kaja Silverman illuminate the poetics of this process by observing how acts of shared memory provide 'the disembodied "wound" with a psychic residence', suggesting that

> to remember other people's memories is to be wounded by their wounds. More precisely, it is to let their struggles, their passions, their pasts, resonate within one's own past and present, and destabilize them.... [I]t is also to enter into a profoundly dialectical relation to the other, whose past one does not live precisely as he or she lived it, but in a way which is informed by one's 'own' recollections.[53]

Silverman is describing the affective impact that Chris Marker's fragmentary screen memories have on viewers of his film *Sans Soleil*.[54] The spectator's role in assimilating the flow of images produced by the film is for Silverman a call to, and an enunciation of, the kinds of recognition that can be produced between remembering and remembered bodies. Like Hirsch, Silverman's model emphasises that any kind of vicarious wounding needs also to recognise the potential for appropriation assumed by its dialectic. That is, it must keep the dialectic internal to the frame that also produces memories as a project of a shared process.

While postmemory highlights the inevitability of feelings that come about through intergenerational memory, the memory affect accounts for how this affective certitude is put into cultural circulation. That is, if postmemory observes the inherited legacies of trauma, then the memory affect emphasises how these legacies are performed as the property and activity of cultural industries. It names the ways in which culture produces *feel-able* pasts about, and on behalf of, others. It here builds on Alison Landsberg's notion of prosthetic culture, which finds that shifts in American screen culture reflect 'a change in what constitutes knowledge', such that the 'cognitive' is 'complement[ed]...with affect, sensuousness and tactility'. Landsberg observes forms that enable an experience 'through which the person sutures himself or herself into a larger history' to take on

> a more personal, deeply felt memory of a past event through which he or she did not live. The resulting prosthetic memory has the ability to shape that person's subjectivity and politics.[55]

For Landsberg, re-experiential engagements with the past have become 'the basis for mediated collective identification and the production of potentially counterhegemonic public spheres'.[56] She thus argues for the benefits of collective knowledge in the place of formerly singular histories. This too involves processes of critical empathy, such that 'the alterity of identification and the necessity of negotiating distances [are]... essential to any ethical relation to the other'.[57]

Contrary to Landsberg, however, the memory affect highlights that there can be no ethical relation to the other that does not recognise that ethics produces the category of that other. This occurs through processes that construct an exchange of feelings, which are often engineered, in simple terms, to produce empathy. Empathy can be understood here, as Ahmed explains, as a ' "wish feeling", in which subjects "feel" something other than what another feels in the very moment of imagining they could feel what another feels'.[58] Lauren Berlant typifies the 'popular belief in the positive workings' of empathy, which operates as a 'rhetoric of promise' when 'the pain of intimate others burns into the conscience of classically privileged national subjects, such that they feel the pain of flawed or denied citizenship as their pain'.[59] This kind of 'sentimental politics', as Berlant calls it, 'confirms the centrality of interpersonal identification and empathy to the vitality and viability of collective life'.[60] In this respect, Landsberg's claims for how prosthetic memories generate empathy neglect to point out that prosthetic culture is deeply invested in the repetition of traumatic affects, the performativity of which is such that 'empathic' subjectivities are often produced as a result.[61] The production of the experiential paradigm in relation to traumatic histories might differently recognise, as Dora Apel has done in relation to Holocaust memory, that many such forms of interactivity contribute to a 'public archive of the Holocaust' constructed of 'always already dead victims'.[62]

Contrary to Landsberg's claims, the memory affect highlights the blind spot in current modellings of loss that position empathy as the morally virtuous way to respond to others. It also points out that this blind spot is a double bind: what might an ethics that begins with a consciousness of its constituting powers look like? How might an 'ethical' relation to the other acknowledge its hidden affective wish – that it constitutes the other from the outset?

For Nicholas Ridout, who writes about theatre aesthetics, a truly ethical project

> would have to confront its spectators or participants with something radically other, something that could not be assimilated by their existing understanding of the ethical. It would have to issue a demand they did not know how to answer. Thus, paradoxically, the value of such an aesthetic production would lie precisely in its not being ethical.[63]

From these perspectives, memory practices that claim to *do* ethics also use it to justify a range of other effects. The wounded other emerges not so much as a project of reciprocal engagement, but as a felt ideal. The memory affect, then, underscores the sentimental politics at work in practices that enable the capacity to feel good about feeling bad about someone else. This is evident in arguments that hold that a cultural apparatus of 'feeling' offers a morally superior outlook by which to cathect traumatic pasts.

Cultural acts that produce contexts for re-feeling the disembodied wound ultimately create ideas about those whose wound it is, and those whose job it is to provide its next point of 'psychic residence'. That is, if the disembodied wound always calls for a new home, then the effects produced by answering that demand still need addressing. Feelings are not merely the result of pre-existing self/other relations, but are the producers and mediators of those relations. This makes clear that the 'loss of loss' still retains embodied effects, which contribute to forms of social memory that may not as yet be recognised as such. The memory affect, then, characterises the cultural reproduction of the lost origin of memory in the sensorial present. It also, importantly, highlights that the staging of the affectivity of the past occludes that staging from the view of those who experience it. These processes occur, as per Berlant, 'when the pain of intimate others *burns* into the conscience of classically privileged national subjects'.[64] In this case, the somatic practices of *burning in* the wounds of others make us sensuously complicit in the creation of wounded and witnessing selves. What Silverman calls the 'new mnemic matrix' of borrowed memories and what Landsberg typifies as the prosthetic 'suturing' of history prompt questions about the kind of experiential repetition that stages itself, experientially, as a 'first'. That is, if memory culture is invested in the loss of loss, then our bodies are implicated in the experiential dimensions of *feeling how we lose it* in profoundly complex ways.

Thinking feelings

Theories of aesthetics, which connect the intersubjectivity of material bodies with the material aspects of culture, open out how relational perspectives of self-making might rethink memory culture's dynamics. As Eve Sedgwick makes clear, touch – as a simultaneous process of sensing, sensation and relation – is invested in such intersubjectivity. Touching produces simultaneous sense-effects and feeling-effects. It is because our haptic senses are involved in the perception of weight, pressure, balance, temperature, vibrations and spatial orientation that they foreground, as Amanda Wise puts it, how 'we receive and perceive the environment' at once 'on the skin and inside our bodies'.[65] We feel *through*, and as, we touch. We feel *about* how we touch at the same moment as we touch. In this way we might develop feelings about feeling: '[a]ffects can be, and are, attached to things...including other affects. Thus, one can be excited by anger, disgusted by shame, or surprised by joy.'[66] Sedgwick's tautology here highlights what Ben Highmore calls 'the cross-modal networks that register links between perception, affect, the senses, and emotions', networks that make it difficult to determine 'a world of touch separable from a world of sight'.[67] Highmore offers 'social aesthetics' as a framework in which 'senses and affect bleed into one another' when 'the sensual world greets the sensate body'.[68] As Jennifer Fisher explains, a refocus on aesthetics can foreground its 'premodern meaning...as having to do with sense perception...as both a performative practice and a morphology of feeling'.[69]

The emergent focus in the humanities on aesthetics as not only an object of enquiry, but also as a method, runs parallel to observations in the social and cognitive sciences around bodily affects, emotions and sensorial processes of self-making. In the sciences, these debates have primarily been considered in studies on the psychodynamics of affect, as well as in psychological and neurobiological studies of emotion, amongst others. These empirical frameworks have informed cultural and philosophical approaches to mind, body and memory, which include anthropologies, phenomenologies and sociologies of the body, and together have fed the more recent turn to affect in the humanities charted above.[70] For my purposes, these perspectives on affect as a mode of memory, as well as memory as an affective mode of self-making, illuminate how the sensorial qualities of memory can be reproduced for some on behalf of others. That is, they explain how the relational processes that underpin all cultural aesthetics undermine the representational paradigms that those aesthetics tend to performatively promise. Affect in this regard, as Felicity Callard and Constantina Papoulias explain, operates not in opposition to memory, but as 'a turn to memory – as long as such memory is understood as embodied and nonrepresentational'.[71]

It is possibly not accidental that with the rise of modern industrialisation and the Freudian psyche, 19th-century psychologists were already preoccupied with the body's ability to produce memory as a sensation of return. Théodule Ribot's observations of affect as the physiological medium of memory underscored what is now a central fascination with the role of feeling in recall, when he determined that

> a really remembered affective state is a present affective state of the same quality, but (at first, at least) of less intensity. It would appear that the emotion arises in memory only when its organic conditions are (not only recalled but) actually reproduced.[72]

While Ribot was invested in a *re*productive account of memory, his contemporary, E.B. Titchener concluded that affect – as the presencing of feeling in the body – *produces* the past. While memory stages itself in the body as a re-sensing of an original event, 'there is absolutely no proof', Titchener wrote, that what is experienced as a 'revived affection' actually is.[73] In this sense, affect is constitutive: it declares the sensory present, which may or may not be derivative of a sensory past. Current cognitive psychology terms this kind of affective upsurge 'episodic' memory as distinct from memories that recall semantics or facts. As John Sutton *et al*. explain, the 'feelings of remembering' created by episodic memory are seen as ' "mental time travel" by theorists who stress the phenomenological reliving of a past event'.[74]

The concept of memory as an affective form of 'time travel' has been most substantively covered by Edward S. Casey, whose phenomenological categorisation of habitual, traumatic and erotic forms of body memory explains that in order for recall to occur, the body provides both 'depth' and 'immanence', in which the past either overwhelms the present or provides it with a potential which may *'become present* in body memories'.[75] Notions of depth and immanence characterise memory's latent grounding of the self in the present, such that '[b]ody memories manifest themselves as continually vanishing into the depths of our corporeal existence – and just as continually welling up from the same depths'.[76] Implied is the idea that there is not only the faculty of sensory recollection but also, equally, the faculty by which the body might sense itself remembering. Here, the act of 'remembering feelings' is accompanied by what Sutton *et al.* call 'feelings of remembering', or what Ribot identified as the body's experiencing of 'a present affective state'. Memory stages itself as a presencing of the past in the body as a process of self-making, and, in Casey's words, 'takes me directly *into* what is being remembered...I leave the heights of contemplative recollection and enter the profundity of my own bodily being.' Memory here becomes a kind of self-travel: 'we allow the past to enter actively into the very present in which our remembering is taking place'.[77]

While for Casey body memory is a storehouse of possible actions and reactions, it also delineates the sensing of that storehouse as an interaction with the lived world. Perspectives that underscore the body's sensing of itself as a process of 'world-making' are key to understanding how memory occurs as a process of felt perception. Thomas Csordas explains that feeling is what mediates the production of the sense of self and its capacity to produce itself and others as objects. This means that, while the senses convey embodied knowledge of the world, they are also formed in relation to that world. They are both perceptual organs and organised properties of a socio-perceptual process. Positioning the body as 'the locus of social practice' places feeling as the continuous perceptual mechanism that at once senses, imprints and is imprinted by culture. Casey explains that such somatic modes of attention are 'culturally elaborated ways of attending to and with one's body in surroundings that include the embodied presence of others'.[78] For C. Nadia Seremetakis, these relations between sense, perception and memory can be understood by viewing memory as an imprint of the socio-cultural process of perception that engaged it. Thus, 'there is no such thing as one moment of perception and then another of memory'. Instead, '[m]nemonic processes are intertwined with the sensory order in such a manner as to render each perception a re-perception... plac[ing] the senses in time and speak[ing] to memory as both meta-sensory capacity and as a sense organ in-it-self'.[79]

In these models, the body can only ever know itself through the knowing that it enables, and is enabled, of itself. Experience, and the way that experience is *experienced* as a self-authored paradigm, is a form of cultural nuance. These ideas are illuminated in the studies conducted by Antonio Damasio, which understand the mind as the constructor of the experience of the self through processes that involve a sensing of emotion. Damasio argues that 'the presence of you is the feeling of what happens when your being is modified by the acts of apprehending something'.[80] One study in particular observes, along the lines of Brennan's thinking, that emotions are physiological as well as psychic actors:

> emotional thoughts, either conscious or nonconscious, can alter the state of the body in characteristic ways, such as by tensing or relaxing the skeletal muscles, or by changing the heart rate. In turn, the bodily sensations of these changes... contribute either consciously or nonconsciously to feelings, which can then influence thought.[81]

Akin to the models of affect put forward by Brennan, Sedgwick and Massumi, Damasio gives science to philosophical perspectives that aim to shift mind–body dualisms, such that 'we only know that we feel an emotion when we sense that emotion is sensed as happening in our organism'.[82] Emotion, then, is a process of sensing a feeling within oneself, which is itself a process of generating a *sense* of self. As I argue in the remainder of this book, memory culture stages this sensing of feeling and its derivatives as a project, and problem, of the traumatic pasts of others.

Sensuous complicities

Given these perspectives on the role of memory in enabling the 'feeling' of the self to occur, affect can describe the means by which memory *presences* itself in the body as a process of self- and world-making. It thereby helps us to theorise the kinds of cultural activities that reproduce the memories of some as an experience for others.[83] Reconceiving memory through notions of affect involves formulating newer models of subjectivity, those which, in Clough's words, 'd[o] not presume loss or a lack in the being in the subject'.[84] Here, the subject positions of wounded and witness do not pre-exist their moment of affective differentiation. Affect is the producer and mediator of difference, not merely a descriptive 'encounter' with it. A shift to a relational framework of cultural memory practices hence allows for the surfaces by which we come to feel each other to be re-felt. Embodying the sensory feelings *of* certain bodies become emotional feelings *about* those bodies: a way of feeling *in* and feeling *about* the world. Corollary of this, we find in Kansteiner's 'consumer' or in Schneider's 'act of labour' a vision of the truth-effects engaged by the tacit sensual reciprocity staged between witness and wounded. An epistemology of affect, for instance, gives attention to the truth-effects of sensorial aesthetics, how they produce meaning, difference and agency, for whom, when and why.

As this book makes clear, performing bodies map the myriad contexts for the affective production of these categories. Chapter 1 begins with the German Holocaust touristscape as an established, even canonical, series of sites that have become the discursive subject of, and enactive of, traumatic affect. It does so to chart the accumulated complexity of the Holocaust affect as a foundational marker of memory cultures and genres. The chapter examines how Dachau Concentration Camp Memorial Site, Christian Boltanksi's *Missing House*, Menashe Kadishman's *Fallen Leaves* and Peter Eisenman's *Memorial to the Murdered Jews of Europe* each map an accretion of the Holocaust affect through producing notions of unspeakability via various forms of spectator engagement. The chapter further analyses the online video work *Dancing Auschwitz* – itself a global media phenomenon – to understand more about how the sites in question manage this particular affect's 'unknowable' social force. *Dancing Auschwitz*, which documents a series of intergenerational acts of dancing at Holocaust memorial sites, interrogates how such sites produce normative behaviours of feeling to enforce the moral and ethical terms by which Holocaust recollection can take place. As an artwork, *Dancing Auschwitz* remediates the Holocaust affect by complicating how feelings of empathy operate alongside the practices of bodily engagement that tend to enforce it. It hence uses *other* feelings to explode the process of 'sentimental politics' in formation; it uses feeling to undo itself. Thus, the work foregrounds how the Holocaust affect operates through spectators to make remembering and remembered subjects culturally distinct at the same time as it contaminates those very bodies it aims to individuate.

Chapter 2 moves from the affective repertoire of Holocaust recollection to the ethno-touristic setting of Australia's Tjapukai Cultural Park, offering an inversion of how the largely site-specific manipulation of the memory affect works. It contemplates the commerce involved in the experiential production of memory culture from the perspective of an unreconciled Indigenous Australia, and in particular, the reconstruction of forms of authentic Indigeneity enacted at one popular tourist destination. Billed as Australia's largest Indigenous cultural park and listed in the *Guinness Book of Records* for producing the country's longest-running stage show, Tjapukai is positioned as experientially and emotionally reconstructive of a repertoire of national cultural memory that uses images of Indigenous Dreaming to whitewash the nation's past. In its reconstruction of the historical corroboree, the memory affect is used to evade and replace the traumas of colonial Australia with a popularised form of cultural memory. I argue that a process of feeling Australian, or feeling white, is brought about through the performative re-enactments of settler/Indigenous relations in the scenarios that Tjapukai creates. It is only when the set script implodes that a very different kind of affective circuitry between non-Indigenous spectators and Indigenous performers is able to be recalled to the present, one that invites a moment of critical awakening that I term 'becoming other-wise'. In becoming other-wise, spectators feel the memory affect's political dimensions. These are those dimensions that hide the ways in which feeling, as an intersubjective process, is used to produce forms of subjective certainty, and here play out in the way that the *sensing* of Indigeneity is itself a 'traditional' means by which settler culture advances its ways of being and feeling normatively superior.

Chapter 3 moves from the site-specific locations of ethno-tourism to the televisually constructed landscape of generic grief performed in the popular psychic television programme *Crossing Over with John Edward*. It builds on the live cultures of memory I have so far discussed to underscore the memory affect as an experience of the traverse across live and televisual spaces of feeling: spaces which are haunted in complex ways by the remediating reciprocities they preproduce for each other. In *Crossing Over*, a complex play with the experiential tropes of trauma, selfhood and grief proliferate as ghostly consumables within new millennial life. This is brought about by the creation of modes of 're-feel-able' trauma for the programme's live studio audience, which are then screened for a home-viewing public. I argue that through *Crossing Over*, the memory affect comes to be shaped as a form of liveness that is built specifically for televisual transmission, denoting the doubled sensory domain of *feeling, remediated*. This explains the idea that memory affects, as formations of liveness, are sensuously contaminated by the re- and premediating principles of the televisual apparatus. The project of feeling memories proposed by *Crossing Over* is hence invested in positioning the primacy of the televisual as an originary sensory source, which, at the same time, performs and produces its own televisual afterlife. I finally suggest that in locating the postmillennial rise of televisual ghosts as one form of memory culture, the specifically topical September 11 ghosts that almost fatefully appear in *John Edward's Crossing Over* exemplify the force of affect in generating publics of feeling around political events. That is, the '9/11' ghosts that he aims – but fails – to mobilise, reveal a *felt* absence of feeling that implodes the normative dimensions involved in everyday ways of feeling the memories of others.

Chapter 4 traces the notion of *feeling, remediated* into the virtual domain of Second Life through a study of the United States Holocaust Memorial Museum's Kristallnacht in Second Life. By connecting virtual Kristallnacht to the material memorial framework charted in Chapter 1, I evaluate the consequences of the digital dematerialisation of memory for spectators, victims and survivors of collective trauma histories. I hence identify the memory affect, not only as it recycles genres of recollection, but also as it repurposes affects of recollection. I consider that, where in Chapter 1 the Holocaust affect was outlined as a specifically material eventuation of the interplay between physical sites and spectator bodies, here, the affective dimensions of material authenticity, which conventionally perform as an unmediated source of sensation and sentiment, are inverted. I consider what becomes of a framework for feeling the return of memory when its primary sensory paradigms are subject to the kinds of undoing enabled by digital interfaces. That is, the memory affect, operating as a mediation of the past that always tries to secrete its mediation in the present, becomes a remediation of that mediation in Second Life. In building on the concept of remediated memory affects outlined in Chapter 3, the memory affect here is hyper-indexical; a cultural formation that constitutes the remembering-avatar-self as *a-material* – neither not a body, nor *not not* a body. I suggest that it is precisely this condition of a-materiality that constitutes the hidden affective indifference that is felt in Second Life; an ontological dissonance that the site works hard to suppress. In identifying those moments in which the avatar-self is able to 'feel' its own a-materiality as a rupture of the otherwise seamless hyper-indexicality Second Life assumes, I present a reading in which the avatar's clumsy awkwardness comes to, in fact, reveal Second Life's greatest political potential.

If chapters 1 and 4 operate as thematic book-ends, mapping correlations in memorial and museum design through the ways that the memory affect can be site-specifically or virtually called up by the body as a process of subject-making, Chapter 5 offers a concluding argument for how the commodifying operations of the memory affect might be meta-affectively undone. With the notion of *affect's spill*, Chapter 5 considers the canonical theatre work of Italian theatre company Socìetas Raffaello Sanzio as a *de-remediation* of the cultural logic by which the spectatorship of difficult histories is medially produced. In doing so, it highlights a specific aesthetic practice that draws into sensory view the 'affectivity' of the memory affect itself. In this, I contemplate a form of *spectatorship that hurts*, a form that enables the feeling of feeling to occur. This feeling of feeling, I argue, is a means to enable the *affective re-perception of affect* as it moves as a specific kind of politics of memory's sociality; one which draws bodies into, and uses bodies to make, those biopolitical domains that regulate and produce cultures of memory. This spill of affect, I suggest, produces an ethics and politics of cultural memory that is largely antidotal to the cultural landscape that the book has elsewhere mapped. Contrary to the commodification of loss demonstrated in the forms of memory culture covered elsewhere, I argue that Socìetas Raffaello Sanzio shift the sociality of the memory affect into a sensory mode that enables the feeling of lost loss to be re-felt.

This moves the book's critique of mass cultural forms of experiential recollection towards forms that use the experiential in a self-aware manner. Affect's spill becomes a sensory occasion that might possibly be closer to Massumi's notion of the pre-social dimensions of affect as a new formation of agency, except here it is identified as tautologous, doubled, meta-affective. Affect's spill describes that sensory instant when affect punctures the social and political dimensions of memory's affectivity.

As the tattoos that open this Introduction suggest, the cultural landscape of lost loss doubles memory's interiority in precarious and anxious ways. By foregrounding a viewpoint of relational bodies in memory, and in understanding that feeling broadens the scopics enabled by vision and is itself performative, empathy is no simple effect that a body experiences in relation to another. Looking for languages that describe the multisensory engagement of bodily faculties used in representations of cultural memory is one way to signal how memory culture demands an 'ethical' response to the project of feeling others and is constitutive of the terms by which those others come to be felt. This form of memory work might then understand that the positioning of 'victim' histories as histories which ought to be felt creates a culture of rememberers who return to the present touched by their own capacities for feeling and, sometimes, not much more.

1
Sensing the Holocaust Affect: Memorials in Repeat, Revision and Return

Figure 1.1 Jane Korman, 'I Will Survive', *Dancing Auschwitz* (2010)
Source: Photo copyright Jane Korman.

Dancing on graves

In January 2010, a curious form of Holocaust art appeared on YouTube. It featured the father of artist Jane Korman, Holocaust survivor Adolek Kohn, dancing to Gloria Gaynor's 1978 disco anthem *I Will Survive* with five of his grandchildren across memorial sites in Central and Eastern Europe. Korman's family covered Auschwitz Death Camp, The Absent Synagogue, Radagast Train Station and Lodz Ghetto in Poland; the Terezin Ghetto Fortress, Theresienstadt Concentration Camp and The Maisel Synagogue in the Czech Republic; and Dachau Concentration Camp in Germany. Their dancing was simple and jovial, their choreographics reminiscent of the kind of dancing that happens at family functions, recalling easy moves, such as a shoulder shimmy or a step-to-the-side. Their attitudes were neither attention seeking nor self-conscious, but rather seemed to rely on familiar practices of family photography. There is no sense, at least in watching the clip, that its performers had intended for it to become an online sensation, earning over 700,000 hits in under two weeks,[1] or that in the ensuing days, the global media would alternately charge it with claims of disrespect or applaud it for its life-affirming vision.[2] In October, *Dancing Auschwitz* won the People's Choice award for the Best European Short Film at the 2010 DokumentART Film Festival in Neubrandenberg, Germany.[3]

In this chapter I examine *Dancing Auschwitz* to consider how iconic memorials situated across Germany's touristscape create opportunities for sensorial and sentimental engagements with the past. In particular, I chart the memory affect in its guise as a *Holocaust affect* – and I build on Vivian M. Patraka's notion of the 'Holocaust performative' to emphasise how histories of the Holocaust are mediated through contexts which enable bodies to feel the pasts of others. Patraka's study of the United States Holocaust Memorial Museum focusses on how the idea of the Holocaust is constructed through gestural interactions with material history.[4] This is cast through the replacement by museum visitors of the Holocaust deceased, where '[i]n a museum of the dead, the critical actors are gone' and it is instead 'the museum-goers...who constitute the live, performing bodies'.[5] The practice is such that we 'rehearse with our bodies...the immeasurability of the loss' and in doing so 'perform acts of reinterpretation'.[6] In these acts, we assume a 'conditional subjectivity', which materialises a 'Holocaust performative' in the interstice between the 'historical real' and the 'live, embodied, disappearing moment of performance'.[7] If the Holocaust performative points to the constitution of the idea of the Holocaust as immeasurable loss, then the Holocaust affect points out that feeling is the medium of the perception of this loss and its constituting message. It is one dimension of how the Holocaust idea is created and is also its designated performative effect.

As I established in this book's introduction, the memory affect is a concept that aims to unpick how culture reproduces the *feel-ability* of memory as a practice of ethical transmission. This argument understands that feeling is produced as a delimiter of difference, rather than as one morally superior form of response to it. In this chapter I argue that it is the specific kind of *feel-ability* produced through, and about, Holocaust history that works to produce it as an 'originary' trauma. This connects to Patraka's observations of the way that the term 'Holocaust' functions as public 'performance'. In it, she observes the 'evocative power' of a linguistic sign, which contains 'all the protocols of the unspeakable... and a sense of unlimited scope to the pain and injustice'.[8] As Naomi Mandel explains, the idea of unspeakability carries an ideological 'emphasis on the limits of thought, language and representation' in recollecting traumatic histories.[9] It foregrounds an impossibility of response – those acts which 'modestly and self-consciously gestur[e] toward [their] own limits' as 'ethical practice'.[10] In other words, unspeakability as a limit point of recollection is also an enactment producing other social effects: it 'reflects a certain self-congratulatory morality... under the guise of not wronging the victims'.[11] The Holocaust affect makes clear that spectator bodies interact with material history to confirm or contest the term 'Holocaust'. Further, this interactive zone of embodied practice, material remnants and replica objects is played out through constructions, remediations and expressions of affect, which also have, and are, material effects.

As one expression of the broader terrain of memory affects covered in this book, the Holocaust affect shows how the tourist body is enmeshed in the production of unspeakability in complex ways. That is, it explains the ends to which performance reinscribes trauma as a limit point of representation. While the Holocaust performative points out that any iteration of the Holocaust as an idea marks the event as singularly unknowable, the Holocaust affect highlights that it is the *staging* of this unknowability as a distinct kind of feeling (the feeling of an absence of feeling, perhaps), which is productive of varying kinds of sentimental certainty: particularly, those which come about as constructions of victim and witness identities.¹² In this way, a work such as *Dancing Auschwitz* might be seen to reveal the 'sentimental politics' of these ideological underpinnings, to recall Lauren Berlant from this book's introduction.¹³ As a result, it critically ruptures the relationship between sentiment and sensation, and instead invites us to re-perceive the framework of that sentimental sensorium. In fact, it isolates the cultural significance of memory's affect and redirects this into a global *elsewhere*, enabling spectators of the work to feel *affect's affective dimensions*. *Dancing Auschwitz* hence shows that if the Holocaust performative produces meanings about the Holocaust, then the Holocaust affect is the felt effect of the truth-effect of those meanings. In it, we embody the feeling of a particular past, and we also embody the idea that this is a past that ought to be re-felt. Richard Crownshaw characterises this as the 'cultivation' of the Holocaust's 'affectiveness', a practice that most often occurs through 'identification with the Holocaust's victims'.¹⁴ Here, unknowability is contradictorily affective, in that it generates the exceeding, but hidden, memory 'charge' that a traumatised victim is given to possess. It is this very cultural practice of *feel-ability* that highlights the assumed capacity for one body's history to be felt by another, as well as for the assumed moral certitude in doing so. In this sense, the Holocaust affect operates to hide the ways in which feeling, as an intersubjective process, is used to produce forms of subjective certainty.

Through a discussion of four very different memorial sites: Dachau Concentration Camp Memorial Site, Christian Boltanski's *Missing House*, Menashe Kadishman's *Shalechet* and Peter Eisenman's *Memorial to the Murdered Jews of Europe*, I examine how culturally scripted behaviours of recollection produce feelings *as*, *about* and *for* Holocaust victims. I further use *Dancing Auschwitz* – an act of intergenerational dancing that occurred at some of the sites in question – to interrogate the function of feeling as it is more usually remediated by them. I argue that the affective engagement enabled by the practitioning of embodiment in such public, site-driven contexts strikes an uncomfortable performativity between *feelings of* and *feelings for*, where tourists are encouraged to perform as 'ethical' subjects whose felt responses to these sites reinscribe hegemonic histories of loss. That is, tourists are co-productive of affect's accumulative power as a mechanism of historical recollection. Their bodies, as sites of feeling, drive and percolate the ascent of its felt force. This is a politics that emerges in the interstices that *Dancing Auschwitz* opens up between repertoires of mourning and their role in producing the continuously performing affectivity of memory sites themselves.

Considering the unspeakable in terms of embodied redoing reveals how bodies not only engage in the cultural sensorium of recollection, but are also constituted by it. Rebecca Schneider points out that all feeling occurs in the body as a kind of repetition, through processes of what she terms 're-gesture, re-affect, re-sensation'.[15] For Schneider, it is the reiterative principle in everyday communicative behaviour that is heightened by 'the explicit *twiceness* of reenactment', which 'trips the otherwise daily condition of repetition into reflexive hyper-drive, expanding the experience into the uncanny'.[16] The notion of *gesturing towards limits* – a literalisation of Mandel's terms – is hence informed by the fact that embodied practices are never entirely singular, but are rather affectively built as an intersubjective phenomenon that accumulates in fixity over time. This is the paradox situated at the heart of memory culture. Its explicit gestural regimes rely on repetition to make concrete and predetermined subject positions a process of phenomenal certainties, certainties that are bound around the sensing of the unspeakable as an exemplary delimiter of intractable difference. *Dancing Auschwitz* makes us see that any gesture towards the limits of representation marks the tourist's encounter with Holocaust history and the Holocaust performative as discursive currency. That is, the positioning of the Holocaust at the limit point of speech also creates subject positions of ethical responsibility in reply. Sensory histories are enfolded into the tension performed between the polarities of repetition and unspeakability, revealing the sticky enmeshment of spectator bodies in any Holocaust iteration.[17]

As the sites I discuss reveal, the Holocaust affect designates the circulation of the embodied capacity to secondarily experience an originary event within mass culture as a unique kind of 'first'. As a presencing in the body, it also importantly hides that secondariness from corporeal view. It is in this way a descriptor for the occurrence of the *loss of loss* as a newly felt event in itself. As a concept, it accounts for the intensities of emotion, sensation or sentiment that arrive through specific forms of corporeal engagement, as well as for the cultural process of circulating that field of sensation. It is hence both the citation of feeling and the feeling of citation: the feeling of feeling as it is transmitted and circulated, as it becomes commodity or even fetish. It produces the sensation of sentiment, and the sentiment in sensation, as political, ideological effect. It folds out of bodies who feel, who perform 'public' feeling, who perform themselves being 'feeling-publics'. The Holocaust affect is not stagnant, nor held by the particular sites that animate it. It is, rather, *moved* by bodies across different sites to mingle with other affective contexts, to get lost, re-routed, short-circuited and hyperbolised. The construction, contagion and accumulation of the felt effects of any encounter with the past come to form the force field in which relations between site and body and body and history unfold.

Embodying unspeakability

My grandfather was imprisoned at Dachau and my journey 18 kilometres outside of Munich city some 60 years after his release is one of many such scenarios of intergenerational return. As one in a group of tourists, I travel to Dachau Concentration Camp Memorial Site by train and then walk in a queue to the gates. Our loose patterning on the entry strip works in counter-imitation to the regulated repertoires that once enforced prisoners to the site. As an act of postmemory, my visit brings into focus the variations on a history that stem from a distant, and difficult, familial legacy.[18] In what ways should I perform the job of cathecting my grandfather's traumatic past? What and how should I feel? Here, the existence of the site as a venue for remembrance seems grounded in a kind of continuous choreographic misrecognition, a staged tension between my movements as a visitor and the site's former repertoires of incarceration. For me, it is the unavoidable indecency of stepping differently that creates an implied otherness, as if my own banal routine makes visible, by negation, the histories held therein. In this, Dachau invites me to corporeally, sensorially invest in the Holocaust as a site of affective difference, which my body then unwittingly mobilises (Figure 1.2).

Sites such as concentration camp memorials, which mourn the public phenomenon of trauma, are secular, liminal spaces that invoke the quotidian performer in heightened engagements of the historically remediated. The politics of feeling held in a site such as Dachau are most strongly produced by the ways in which it recreates lived

Figure 1.2 Nandor Glid, *International Monument* (1968) at Dachau Memorial Concentration Camp
Source: Photo copyright the author.

trauma and stages itself as its aftermath. Dachau, like many such sites, aims to recreate a nominal *sensing* of the presence of the historical event by performing as the future of its past. Dachau hence draws bodies into its dramaturgies of authenticity by acting as both a metonym of the Holocaust and a material index of its legacy. It stands in for the Holocaust, and via its performance of remains, is also tangible evidence of its history. As Barbara Kirshenblatt-Gimblett has explained, such 'in situ' sites rely on an 'art of mimesis' by expanding the world of the ethnographic fragment to entire representational displays.[19] In memorial concentration camp sites, however, it is an expanded 'poetics of detachment' that informs the in situ landscape inhabited by tourists.[20] Visitors travel through a post-traumatic after-effect. Through this staged materiality, spectators are invited to sensorially invest in an imaginative reconstruction of the original site, as well as to complete the image of loss that that original site marks. Here emerges the central torsion staged by the site: bodies experientially 'sense' the unspeakable, to also become referents of it.

Performance theorists have long debated the role of the body in the transmission of cultural memory. These are ideas that most firmly coalesce around notions of repetition, evidenced in what Rebecca Schneider terms 'sedimented acts', and converse notions of the ephemeral, typified in what Peggy Phelan calls an *'active* vanishing'.[21] For Phelan, bodies move through time in a way that stages loss. Whether this loss is the tangible loss of a relative or the intangible force field of social relations, the doing of loss is a determining process of identity:

> Identity is perceptible only through a relation to an other – which is to say, it is a form of both resisting and claiming the other, declaring the boundary where the self diverges from and merges with the other. In that declaration of identity and identification, there is always loss, the loss of not-being the other and yet remaining dependent on that other for self-seeing, self-being.[22]

As Phelan has made clear, the ontology of disappearance enabled by the live performing body is evidence of its capacity to operate beyond representation.[23] What performance 'loses' by way of its 'drama of corporeality' is central to the contingencies of the knowledges it constructs.[24] There is an implicit phenomenology in Phelan's model: loss is the process of *feeling* by which subjectivity works.

Phelan's model has been seen as distinct from ethnographic models of performance, which emphasise memory's transmission over its disappearance, its capacity for contest, accumulation and revision, over loss. Paul Stoller, for instance, observes in the spirit possession practices of the West African Hauka that ritual embodiments constitute a relation between 'the centrality of the sentient body' and political power, such that the body becomes an 'arena in which cultural memory is...refashioned'.[25] Joseph Roach and Diana Taylor observe similar patterns for the preservation of cultural memory in the West. Roach argues that practices of surrogacy retain knowledge across time: a substitute can only stand in for what was there by pretending that they were always there.[26] For Taylor, substitution is one aspect of 'embodied memory:...all those acts usually thought of as ephemeral, nonreproducible knowledge'.[27] This involves recognising that the body transmits 'knowledge and memory... [as] a product of certain taxonomic, disciplinary, and mnemonic systems'.[28] In this, any action is deeply recollective, repeating codes (or 'hacking' them) across time. Schneider sees the space for bodily knowledge in the 'affective transmissions of showing and telling', the ' "sedimented acts"... that haunt material in constant collective interaction, in constellation, in transmutation'.[29] Here, re-feeling can operate as a kind of 'empathy' in which the coalescence between an embodied action and its cultural signification is prosthetically understood while being produced.[30]

The notion of the Holocaust affect as the felt effect of the truth-effect of Holocaustness indicates how the present gestural behaviours that occur at sites of recollection are inflected with past practices (past tourist practices, which stand in place of the past practices of the deceased), and intimates how memory culture relies upon live bodies at large. In memorials, it is the material body of the tourist that remediates the absent bodies that comprise the scene of trauma. Liveness is staged as a recall of those who have disappeared. Indeed, one could argue that a viewpoint of the quotidian performer as always productive of ontological loss feeds the framework by which loss is discursively repeated, such that the 'drama of corporeality' is what becomes gesturally routine across touristic time. Here, the *modus operandi* of trauma memorials conflates the durational, material experience of bodies that were lost, with the durational, material experience of bodies given to recollect loss. In this we understand that if in everyday life we embody the gestural habits of the past as surrogates, in trauma memorials we embody a site's consciousness of this very capacity. In doing so, we stage meanings about our role in re-embodying the histories of others. Performance not only entertains unspeakability as a paradigm of representational concern, it produces the discourse that creates it.

Considering the unspeakable in terms of gestural repetition reveals how spectator bodies engage discourses of recollection in order to produce other outcomes. Between the perspectives put by Phelan and others, we understand that spectators are given opportunities to feel, and feel about, the past in a process that performs the non-recuperability of memory while also transmitting it. Public memorial sites in particular use the sentient body to occlude how embodiment can produce, rather than merely fail at, these kinds of transmissions. In this, Holocaust memorials might be interpreted as creating a repertoire of unspeakability that is produced through affective stagings of the past in the present. That is, the hidden phenomenology of Phelan's model reveals the constitutive functions of embodied practices: the unrepresentability of trauma is produced by spectators as a kind of felt ideal that *then* becomes productive of spectator/subjectivity certainties. Here, the 'feeling' in Phelan's invocation of 'not-being' but 'self-seeing' is one that *does not result from*, but one that *produces*, meanings around the very 'being' that one is 'not'. To paraphrase Geraldine Harris: if the notion of unspeakability is constitutive of the 'Real' that representation is not, then representation is only ever constitutive of that which it fails to be.[31] A repertoire that performs the unspeakable through varying forms of affective engagement fails to acknowledge that there is nothing outside of the act that constitutes the meanings, or affects, which create it.

Staging the affect-effect

At Dachau, tourists are invited to complete linkages between moments of feeling *like* a Holocaust victim, and moments of feeling *for* that victim. These links are encouraged by various site-specific invitations towards the affective repetition of loss that the site offers. Inside the site, we wander the grounds. We watch footage of the camp, filmed by the Allied Forces, to educate us about its atrocities. We walk through bunkers which illustrate the squalid lived situations of their former inhabitants, but which are themselves replicas, producing a kind of deferred affective contagion (we are told that the originals were destroyed because they were contaminated with disease). Contaminated, we are, by the imagined contamination carried by bare wood replica beds, we are next invited to step into the crematorium (not an imitation) as if here, our hypothetical contagion might be burnt away. The mostly unadorned external landscape, the factual museum display boards, the abstracted external monuments, the religious memorial sites and the documentary film all work together to produce the 'punchline' that is to land in our bodies with the moment of stepping into the crematorium. And it is here, at the wound point, that tourists (like me) *mis*-behave. On the day I visited, I chose not to step inside the drum. Others stood in damp, sombre silence, but most, on the day of my visit, took photographs.

Teresa Brennan explains that affect can be understood as the physiological shift that accompanies 'judgement' – where judgement is any evaluative relation towards or away from an object. The experience of affect resides in what might be considered its derivatives: 'mood', 'emotion' or 'sentiment'.[32] For Brennan, 'all affects...are material, physiological things'.[33] In Brennan's study, the transmission of affect can be explained in the neurology of smell, for example, in which

> pheromones – molecules that can be airborne and that communicate chemical information – signal and produce reactions by unnoticeable odor in many hormonal interactions, including aggression as well as sex...smell...is critical in how we 'feel the atmosphere' or how we pick up on or react to another's depression.[34]

Further, 'the pheromonal odors of the one may change the mood of the other...so-called interaction changes our biology'.[35] In understanding affects as material entities that impinge upon sensing bodies, affect describes relationality at work. While one's 'mood' is experienced as autonomous, the passage of pheromones makes clear that embodied states are intersubjective. *My* sentimental disposition and the imprints by which *I* feel it are not only in response *to* you, but *are* you. I ingest something of you that will materially change me, and this will cause changes in another again.

For Mark Dapin, Dachau meets an urge to experience *the experience* of Holocaust victims: '[P]eople want to know what it looks like, what it feels like. They want to know what a Jew feels in a concentration camp.'[36] Dachau offers the promise of an intercorporeal (or 'post-corporeal') exchange between visitor and victim subjectivities as part of its dramaturgical logic. That is, it stages the history of victim identities as an affective residue located at the site, one that pre-exists the spectator bodies who then experientially source it:

> everyday sights become sinister: the slow raising and lowering of the boom gate; a man wearing a heavy coat collecting logs in the

snow; smoke climbing from the chimneys of the houses that abut the fence of the camp.[37]

As the Dachau website explains, the grounds aim to teach visitors factual history, and then encourage them to imaginatively explore that history in lived memorial space:

> One should plan to spend the first half of the visit touring the main exhibition in the former maintenance building...The rest of the time can be spent viewing the grounds and the supplementary exhibitions in the bunker, the model barrack, and the crematorium.[38]

This demand for feeling the 'post-corporeal' staging of the experiences of victims is what establishes the invitation to form 'judgement' out of the sensory mode of 'feeling the atmosphere'. Judgement emerges as the formation of emotion in relation to the heightened sensory awareness, evoked, for instance, by stepping into the crematorium. Affect is the sensorial register produced by the death machine and the feelings, images and words transmitted by other tourists as they corporeally negotiate it.

The intersubjectivity of affect staged at a site such as Dachau holds implications for how we consider the role of feelings in creating cultures of memory. If affect is relationality at work, and if this form of 'prosthetic suturing' produces an ethical relation to the history of the other, then what affect promises and how it is practiced are enfolded in constant, hidden tension. While visitors can only ever corporeally *mis*recognise the site's historical repertoire, it nonetheless aims to produce a forceful alignment between feeling *like* a Jew and feeling *for* Jewish victims. What it doesn't make clear is that this very promise of identification involves, as per Brennan, a sensing of Jewish victims as it is produced in relation to a sensing of visitor selves. The formation of an attitudinal judgement through smell, touch or material interaction is invested in the production of the *other as a sensed process*. Stepping into the drum, therefore, involves the creation of victim identities through the sensing of oneself as distinct from the wounded other, who occurs as a kind of 'felt ideal'. This reading pushes accounts of the unrepresentability of loss to theorise the permeability of body to body, and body to environment, that this theory hides. Arguments that emphasise performance's contingency as vital to the knowledges it creates also see that performance 'remains' complicit in the production of those notions that come to mark it. The construction of the affects of a wounded people instead become appellative in extension of Dora Apel's words, as I noted in my introduction to this book: they not only figure Jewish people as always already dead victims, but also position tourist-visitors as always already their saviours.[39]

This creation of points of identification through staging affect connects to what Gary Weissman has termed a 'fantasy of witnessing' operative in post-Holocaust culture, which emphasises 'that the experience of listening to, reading, or viewing witness testimony is substantially unlike the experience of victimization'.[40] For 'nonwitnesses', as Weissman explains,

> one's own place in the hierarchy of suffering has much to do with one's professed ability to 'feel the horror.' One's intellect and moral fiber are measured by the degree that one has come... to 'endure the psychic imprint of the trauma.'[41]

As Weissman makes clear, the desire to experience Holocaust suffering is 'the kind of wish only one who was *not* there could have, and only then because one knows it cannot be fulfilled'.[42] It is, however, the moral performativity produced by nonwitnessing which is important: ' "trauma" becomes the sign of one's authentic relationship to the Holocaust'.[43] Mobilising the sign of trauma as that which can be momentarily re-felt involves a politics in the site's use of affect to encourage identification with what Omer Bartov describes as historically binding 'German self-perceptions and attitudes toward Jews' and 'Jewish self-perceptions and attitudes toward real and perceived enemies'. As Bartov explains, both perspectives enfold 'numerous connections between the German discourse on nationalism, identity, and Nazism, and the Jewish discourse on identity, Zionism, and the Holocaust' – a discourse on enemies and victims which risks perpetuating the kinds of power upon which the Fascist history first relied.[44]

An ethico-political critique of the Holocaust affect explains not only how tourists engage with the aesthetics of remembrance, but also how they are invited to sensorially produce a self-identifying subject position through complicity with a site's programme. This involves observing, in the words of Dominic LaCapra, that any perception of victimhood should first recognise oneself in Himmler.[45] It is through strategies of feeling like and for Holocaust victims that Dachau encourages a form of nonwitnessing that, to re-cite Mandel, can also produce a kind of 'self-congratulatory morality...under the guise of not wronging the victims'.[46] If the nonwitness desires the subject status of victim, and if this desire takes the shape of 'empathy' as a 'wish feeling' in Sara Ahmed's terms, then it also produces sentiments which, as Weissman attests, 'may impede consideration of one's potential to occupy the position of perpetrator or bystander'.[47] The crematorium entry invites visitors to construct the affective realm of the trauma of others, and in so doing, enables visitors to affirm their own domain of moral self-certainty. If tourists only ever experience Dachau as a site at which the trauma of others is perceived to have been caused by *other others*, then the site's promise of 'never-againness' will only ever mark a crime that operates beyond what the visitor is able to perceive of themselves. A politics of affective identification thus understands that memory culture performs the feel-ability of others as equally as that feel-ability is performative. That is, feelings reflect and are productive of specific relations of power.

Syncopated histories

I walk along Berlin's Grosshamburger Strasse and I miss it. I walk back again and still miss it. There are no signposts to this memorial, but I know that it exists. I walk again and realise that what I continue to miss – Christian Boltanski's *Missing House* – is supposed to be missed. When I finally see it, I see its *absence*. This artwork is site-specific, tucked away on a narrow strip, which is gentrified with cafés and high-end fashion shops. Amidst the daily bustle of urban dwellings, the artwork marks a moment in history: on 3 February 1945, Block 15–16 Grosshamburger Strasse was destroyed in Allied bombings as part of a raid that de-housed over 120,000 civilians. As Abigail Solomon-Godeau explains, in this act of destruction

> [t]wo of the building's central staircases and the apartments on their landings were destroyed; those people in those apartments on those stairwells were perhaps maimed or killed; those living on either side were spared.[48]

In perceiving the house's absence I trespass on private property. The missing house is small and discreet: gated, sitting in a landscaped garden that bridges apartments on either side. Close-up, the house is suggested by mounted plaques on a neighbouring façade, which list its former inhabitants, their professions and their periods of tenancy.

Boltanski has said that his 'work is not about, it is after'.[49] Beholding the house is to behold its *dénouement*. In order to see it, I need to reassemble its absence. This is a simultaneous act of resurrection and destruction. I am made complicit in the act of absence-making and I have no choice in my subject position: I am guilty of making the work *work*, which is also to say I am guilty of making the house re-disappear. Boltanski's art, like Bartov's scholarship, reconsiders the function of victim–perpetrator positions in contemporary representations of traumatic history. However difficult, in the post-Holocaust contexts of Germany and Poland, an intersubjective approach can, as Steven L. Sampson observes, mitigate questions of culpability to generate 'new "conversations"' and '[a]lternative versions of truth'.[50] Contrary to the subject positions set up by Dachau, *Missing House* reinscribes and complicates victim–perpetrator polarities by inviting us to *re-feel* the process of afterwardsness that it stages. Here, Boltanski materialises a violent history and also a perception of outsiderness to it. These multiple points of identification become moments of perceptual reckoning: my reproduction of the house does not place me inside it, but places me as an executor of the violence enacted upon it. My annihilation of the work is central to my ability to read it: I do not merely peruse the life of the victim as I do at a site such as Dachau; I contribute to their immanent destruction (Figure 1.3).

The sensory complexity highlighted by *Missing House* draws Brennan's science of affect towards its philosophy, to argue that what affect stages is not only feeling, but a 'bodily encounter' – as Susan Best puts it – a felt corporeal 'syncope'.[51] In Massumi's writings, this syncope is explained by affect's 'two-sidedness': the 'participation of the virtual in the actual and the actual in the virtual, as one arises from and returns to the other'.[52] Where Brennan differentiates between physiology and the judgements of mood or emotion that accompany it, Massumi differentiates between what he calls potential and its expression, understanding emotion as one actualisation of a possibility, or the 'point of insertion of intensity into semantically and semiotically formed progressions'.[53] If emotions are the

Sensing the Holocaust Affect 43

Figure 1.3 *Missing House*, Christian Boltanski (1990)
Source: Photo copyright the author.

bodily expression of affect, then the moment of syncope put forth by Best designates the precise bodily pause during which a form of 'pre-expression' is sustained: it is 'the turning point at which a physical system paradoxically embodies multiple and normally mutually exclusive potentials'.[54] In Brennan's terms, this syncope might detail atmosphere entering the individual alongside the uncanny sense that it has done so. Here, processes of feeling while also *feeling feeling's transmission* engage meta-sensory mechanisms, such that, as Massumi cites of Spinoza, affect designates '[...an impingement upon] the body, and *at the same time the idea of the affection*'.[55] One potential of affect then is that the body *in potential* might feel itself feeling itself coming into being and meaning.

For Clare Hemmings, exponents of affect such as Massumi tend to position it as an antidote to arguments that we are all catastrophically 'caught in culture'.[56] And yet, as Brennan makes clear, affect is the biologically social and sensorial hiding of that important duality – while 'people experience themselves as containing their own emotions', the 'psychosocial actually gets into the flesh'. In sum, 'affect is a process that is social in origin but biological and physical in effect'.[57] The fleshly nature of the social is exemplified in what Hemmings sees as the 'delights of consumerism, feelings of belonging attending fundamentalism or fascism', which are all 'affective responses that strengthen rather than challenge a dominant social order'.[58] In these examples, the body's experiencing of itself as containing its own feelings might rather be seen as evidence of the coercive uses to which affect is put. Here, what Jeff Pruchnic and Kim Lacey call the 'rhetorical' functions of affect become clear. In this model, present sensory experiences are informed by 'existing pre-conscious memories, particularly [the] recollection of images and marketing messages and the somatic markers created by those commercial appeals'.[59] In studies of the effect of advertising on sensory perception, Pruchnic and Lacey, for instance, point out that:

> When we have an affective response to an object or concept, it becomes 'tagged' or 'marked' in our memory with that feeling, increasing our chances of responding in similar fashion when confronted with or recalling the original stimulus as well as others that we perceive.[60]

In cultures of memory, the kinds of aesthetic strategies that either use or challenge affect's 'somatic markings' reveal the rhetorical functions invested in processes of feeling for, and feeling about, others. It is in this way that the strategies of making disappearance re-disappear mobilised by *Missing House* can be seen to foreground, as well as complicate, how affect operates, rhetorically, as memory.

Missing House overturns the straightforward alignment with victim identities produced at a site such as Dachau by instead posing problems for corporeal processes of recollecting loss. As a spectator, I feel my perception of re-disappearance while I also perceptually produce it. The memorial sustains present and past: I am at once within and outside of historical sequence. The duality of these unfolding times enables time to permeate time, holding me in chronotopic suspension. Like a feedback loop, I become the linkage point between pre- and post-trauma. In becoming the channel through which time can bleed, I am the conduit by which feeling is self-transmitted. Affect becomes the transmission of feeling and the feeling of the process of its transmission. In seeing destruction at the same moment that I reconstruct it, the loss of the house that I perceive is short-circuited by how I perceive my co-creation of it. My position as a mourner of loss is punctured by my immediate complicity in crafting it.

While the Holocaust affect in Boltanski's house is wrought by a pre-/post-temporal complexity that engages the spectator in a process of feeling their own looking – a 'two-sidedness' or 'participation of the virtual in the actual and the actual in the virtual'[61] – this sensorial enmeshment is made further problematic by the house's hidden history. As Solomon-Godeau explains, the house carries multiple histories of trauma:

> prior to 1942 many of the building's residents were Jews. By the time of the bombings, however, those tenants had been evicted, displaced, deported, and presumably liquidated. Thus, when the allied bombings occurred, many of the tenants were German Aryans who had replaced the now-vanished Jewish residents.[62]

As a memorial to the lives of Aryan-Germans reinhabiting the site of already 'liquidated' Jewish-Germans, *Missing House* charts an uncomfortable sense of double death. On one hand, it mourns Aryan and Jewish subjects and at the same time it positions Allied Forces and Nazis as perpetrators. This is a marked point of difference from the generality of memorials that focus on Jewish-German perspectives alone. On the other hand, in positioning spectators as the perpetrators of Aryan-German deaths, it upsets the conventional framework of empathy produced by a site such as Dachau. Here, it enacts a kind of surreptitious death wish towards the Germans it would rather mourn.

There is hence a strange afterlife to Boltanski's *Missing House*, revealing the Holocaust affect as that which attempts complex relations to spectatorial productions of history and that which cannot escape the affective complexities carried by history itself. In doing so, the work foregrounds the ways that bodies are sensorially, emotionally 'tagged' by the force of the Holocaust affect and also become its 'taggers'. At the same time, the rhetorical problem of affect can also be found in the work's deployment of the trauma of the Second World War as a canonical genre – there are no facts that detail the surreptitious history the house hides. In this way, Boltanski constructs a memorial that is to some degree exploitative of the traumatic tropology of Berlin itself; in his own words, by 'tak[ing] any house in Paris, New York, or Berlin...you can reconstruct an entire historical situation'.[63] Given the particular historical weight of the house, the absence that I am forced to remake as spectator masks the equally sinister absence of its originary Jewish inhabitants. These inhabitants are made doubly absent for being un-recollected by the memorial. While the Holocaust affect is engaged here in a complex shifting of spectator identifications, that shifting is caught up in another kind of erasure. I become complicit not only in the destruction of the Aryan inhabitants, but also complicit in the surreptitious 'secret' surrounding the lost Jewish bodies that the house seems to maintain.

Accumulating affect

In the Memory Void of Berlin's Jewish Museum is the installation *Shalechet* (*Fallen Leaves*). In it, sculptor Menashe Kadishman invites visitors to tread across a bed of screaming iron faces – faces that are reminiscent of Edvard Munch's *The Scream*, but that are made more horrific for their uniform anonymity. Over 10,000 'open-mouthed' and 'coarsely cut' faces flank the Void's floor to demarcate a pool of silently wailing objects.[64] If Boltanski's house grounds the spectator in a practice of complicit destruction framed by relations between site and sight, then Kadishman's faces nullify the validity of sight altogether, engaging a practice of sounding, listening and sensory moving. Where Boltanksi's house focalises absence in order to mark (and mask) histories of erasure, Kadishman's faces envision the very shriek of destruction his absent house might otherwise sound. In Munch's *The Scream*, a fire-red sky circles a lone body, its head held between two arms, looking out in horror at the so-called 'scream' of nature.[65] Fredric Jameson has reflected on the decay of

Sensing the Holocaust Affect 47

Figure 1.4 Menashe Kadishman, Memory Void Installation *Shalechet* (view to the passageway), (2001)
Source: Copyright Jewish Museum Berlin. Photo Marion Roßner.

the self-possessing modern subject depicted by Munch, to argue that Munch underscores a 'waning of affect' in an image that fails to express vocality through paint.[66] In Jameson's reading, *The Scream* is 'an embodiment not merely of the expression of... affect but... a virtual deconstruction of the very aesthetic of expression itself'.[67] If the paint in Munch's *The Scream* embodies the human in crisis, then Kadishman's faces stage the crisis of humanity itself (Figure 1.4).

In Silvan Tomkins' studies on affect, the face is a primary medium of interpersonal affective communication, exposing links between the imitative origins of externalised gesture and the alternately internal experience of emotion. Contrary to the psychodynamics of affect theorised by Brennan, Tomkins' theory observes the role and participation of the visual in a host of other felt experiences or drives:

> Since it is known that the smile of the face of another is a specific activator of the smile of the one who sees it, the awareness of the smile in the self may release another smile either on the basis of the similarity of the smile in the visual and the smile in the proprioceptive modality, or on a learned basis, since one's own smile was often preceded by the smile of another.[68]

Tomkins' reading of the ways in which interpersonal exchange can precede autonomous physiological processes positions the face as the expression of an internal state (that is, as an expression of a felt affect) and as an affective originator, whereby, in being triggered by the smile of another, physiology can precede (or cause) a forthcoming point of emotion. This system envisages, in the words of Anna Gibbs, an unbeknownst process of 'affect contagion'[69] in which bodies communicate and even generate an exchange of feelings in ways they do not realise, or more importantly, *feel*.

Kadishman's faces sculpturally re-imagine the contagion of affect that the living face can produce, and they do so by mouthing silent shrieks of horror. While, as Tomkins points out, it is the visuality of the smile that transmits affect's movement, Kadishman's faces redirect the body's sensitivity to the role of the face in making affect move. In visualising shrieking, they send out a silent scream: swallowing their sound, and sounding the silence they emit; their faces register an inability to vocalise and we feel their non-speaking reverberate. This occurs in the invitation for visitors to step on the faces' non-speech. When walked on, the faces clank, evoking the clatter of death trains or a chorus of heavy bells. Emil Hrvatin has argued that in the act of the scream '[t]he breakdown of the subject has already occurred...the scream relates the condition of the subject for whom help always arrives too late'.[70] Hrvatin references Slavov Zizek's reading of Jacques Lacan, who 'determines the *object small a* as the bone that got stuck in the subject's throat'. A scream, thereby, is a 'voice that cannot...enter the dimension of subjectivity'.[71] The negated subjectivity occasioned by the scream is made palpable in

Kadishman's faces, which *sound themselves* through the bodies that walk over them. While this involves the lending of subjectivity to those who have lost it, the relation between face and tourist is also the act that occasions their cry.

As in Taylor and Roach's models of performance, *Shalechet* marks the unspeakable as the repeatable – the improvised steps of tourists who corporeally cathect the installation's nod to impossible representation. In this way, Kadishman's faces position us as subjects of Holocaust history and as meta-discursive imprints of it. This occurs in how they stage unspeakability and stage the tourist as its co-scripter. The contagion animated by the installation means that tourists are forced to cathect and destroy the trauma signifier they also seek to witness. As the bone that gets stuck in the victim's throat, help, in the form of a witness, will always arrive here too late. Kadishman's faces make clear that the Holocaust affect is incurred by bodily transmission and is also exposed to exploitation of that transmission. In this sense, it becomes not only the steely shudder of the shriek as it travels up the legs of those who tread, but also the unspeakable itself, which is only ever animated, and felt, by those who desire it. If in Boltanski's house I am unwittingly complicit, here I am confronted by my own willingness to co-create the Holocaust as a site of feeling. This articulates the spectatorial moment in which I feel my desire to feel it. The installation is in this way mnemonic *and* meta-mnemonic. It contemplates the perceptual role of the spectator in determining the 'self-congratulatory' affects that accompany discourses of loss.

Ambulant affects

If the Holocaust affect has so far appeared as the production of the unspeakable in a clammy crematorium, or as the sensorial arrest invoked by envisioning a moment of re-deconstruction, or even as the clanging shriek sent up legs and bodies, then it reappears in Berlin's most controversial Holocaust monument, Peter Eisenman's *Memorial to the Murdered Jews of Europe*, as its own remediation. Here, the Holocaust affect does not invoke polarities of victim, witness or bystander. It entertains the question of speaking, and of not speaking, of unspeakability and its felt unconscious through attention to a focalised, non-deictic present. This monument, which, after over ten years of design and decision making opened in 2005, only refers

Figure 1.5 Memorial to the Murdered Jews of Europe, Peter Eisenman (2005)
Source: Photo copyright the author.

to those who inhabit it. Or, as some writers have noted, it focalises the *felt* nature of the unspeakable to stage 'embarrassed space', 'nothing more than feelings' and practices of 'mutual observation' (Figure 1.5).[72]

Memorial to the Murdered Jews of Europe was designed for remembrance, but it also 'raise[s] the question of remembrance and its location' through 'avoiding creating an impression of finality'.[73] The memorial is disturbing for its resistance of finality on a number of levels. A question put to our tour guide on approach: 'Are they supposed to resemble coffins?' They are coffin-like, the 2711 stern black rectangular boxes rising into the sky at different heights from a grid platform. The measurements of the stelae are precise, forming endless corridors in a paradoxically predictable and yet maze-like formation. The boxes ascend to create waves of undulation as they move up and down again. Their pattern is mesmeric and their presence awesome, expanding over almost five acres of Berlin's Potsdamer Platz. From a distance they consume a whole neighbourhood, writing a blanket of grey disturbance across the rhythms of daily life. Eisenman explains that the memorial 'attempts to decontextualise the Holocaust... Not to try and locate it, not to try and make it a thing of nostalgia, not to try and make it be able to be rationalized.'[74] As one online blogger has commented: 'I'm here. In Berlin. Spent 5 hours at the Memorial yesterday. It's wonderful – because it's the only place in Berlin where I did NOT think of the Holocaust.'[75] In this sense, the stones are a creatively inhabitable civic space: 'I like to think that people will use it for short cuts, as an everyday experience, not as a holy place'.[76] Children play and jump in this strange landscape, however formidable the sharp cement edges may be.

The memorial arrived in Berlin's heightened memory boom as a point of controversy and as a distinct intervention into what the catalogue terms 'Berlin's new satisfaction with itself':

> The *Memorial*... intends to disturb and discompose. Before the city and country lean back and act as if things were back to normal – could ever be back to normal – the question is raised here, cast in stone: what does the city have in mind as it re-moves the ruins of history and makes itself a cosy sitting room again?[77]

The catalogue argues that the memorial 'draw[s] a line in stone under the past'[78] in a way that suggests it draws a line under how memory should function. But 'what message could an abstract monument not associated with a particular place or event be expected to have?'[79] What is it to inhabit, to peruse or eat lunch on a space that insists upon these varying degrees of ambulance? How does this memorial reframe and remediate the Holocaust affect as it has so far accumulated across Berlin's memorial touristscape? What is it to 'lose' the 'loss' that memory would otherwise keep intact?

Ideas of staging the impossibility of traumatic representation have been foregrounded by James E. Young's notion of the counter-monument, which focusses on the ways that design can erase traditional polemics to instead offer performative play between site, rememberer and history.[80] As already exemplified in Kadishman's shrieking faces and Boltanksi's house, the counter-monument is self-ambivalent, emphasising spectator engagement by recognising that it is 'activity that brings monuments into being... [the] ongoing exchange between people and their historical markers'.[81] Lisa Saltzman, in a move beyond Young, investigates the 'postindexical' as an alternate practice of post-Holocaust visual culture. While the index of modernity depended on material contiguity to an original, the postindexical is 'empty' or 'impotent', 'the index at one remove, the index that is no longer a sign, but instead, pure signifier'.[82] Saltzman discusses United States–based artist Ann Hamilton's 1997 *welle*, a white wall drilled with 400 minute holes, weeping water through a gravity-fed IV drip system. The work produces the effect of 'allowing a single "tear" to emerge and slowly gather enough liquid to form a drop',[83] and Saltzman notes how it displaces affect from its source:

> rather than using a set of visual strategies to produce an affective response in its spectator... it is a work that fully embodies or performs affect in an entirely self-enclosed manner... pure affect as visual effect.[84]

While Hamilton's *welle* performs the role of the spectator on their behalf, Eisenman's structure positions the spectator as its memorial. It is in this way not the reconstruction of a lost object but an inhabitable landscape that signifies into the affective construction of loss and its material histories. As such, it forms a postindexical structure that continuously disappears its lost object from the gestural and affective play that it sets up, recalling the spectator as a remembering body to the scene of its role in historical recollection. For memory to work here it resists nostalgic longing and instead refuses what it is most used to naming. Memorialisation is enacted as a statement against prolonging longing itself.

The memorial suggests that it is precisely inside the moment in which the index is refused that another kind of memory emerges: one built out of improvisational gestures given in how the body informs, moves through and writes the space. In this landscape of tiered and blankly ominous boxes, I am caught within my desire to re-experience the loss of another. I am also affecting, such that the memory network built by any cluster of spectators who walk together shares that which won't be spoken by the site. In its focus on the act of walking through space, it reveals that the Holocaust affect is not stagnant, but is *moved* by bodies across different memorial sites. Through it, a practice of feeling the feelings of others meta-affectively emerges as a key *modus operandi* of the public remembrance of trauma.

Meta-affects

Figure 1.6 Jane Korman, 'Art Must Go On', *Dancing Auschwitz* (2010)
Source: Photo copyright Jane Korman.

In view of the affective performativity of the Holocaust staged in *Memorial to the Murdered Jews of Europe*, *Dancing Auschwitz* similarly materialises concerns that meet the discursive intersections of memory and performance in the current moment. Like Eisenman's memorial, in positioning survivors and their offspring dancing on sites of trauma, it challenges accepted practices of mourning and remembrance, and in so doing, unrests the foundational positioning of the Holocaust – or indeed, the normative gestures of the 'Holocaust performative' – from within memory studies' canon. Condemned by some critics for instituting desecration by 'dancing on graves', the work also questions the Holocaust (and more broadly, trauma) tourist industries by inserting into its repertoires of mourning divergent practices of memorialisation. As such, it makes clear how practices of affective engagement with traumatic pasts circulate the unspeakability of the Holocaust to produce spectator subjectivities that are problematically invested in feeling otherness as a statement towards a moot-point practice of ethical recollection.

Jane Korman herself notes that, rather than desecration, the divergent practice of 'dancing' Auschwitz was conceived out of a necessity for her own family members to acknowledge the experiences of their grandfather as well as their relation to an otherwise distant history:

> My intention is to present a fresh perspective to younger generations who have often become numbed and desensitized to the horrors of the Holocaust and other genocides happening now throughout the world.[85]

Her father has described his return to the camp with incredulity: 'If someone would tell me here, then, that I would come sixty something three years later with my grandchildren, so I'd say, "What you talking about?" This is really a historical moment.'[86] In view of how repertoires of mourning generate the Holocaust affect at sites such as Dachau, *Dancing Auschwitz* can be seen to disrupt the link produced by the spectator between the site's affective 'doom' and the ensuing space of moral responsibility made for its visitors. Indeed, in treading on graves, *Dancing Auschwitz* makes what might have otherwise been the unconscious contradictions of smiling while mourning, of photography of the sacred, of indecent gestural codes, visible and empowering. It recodifies the generational inheritance of trauma as a complex, ambivalent and playful, even possibly joyful, undertaking. It traces the Holocaust affect into counter- and meta-affective moves, moods and poses.

If the Holocaust affect moves through us by processes of facial mimicry, the bodily contagion of scent, the role of sight in producing the not-to-be-seen, the role of sound in producing the unspeaking, then Jane Korman's family who 'dance' Auschwitz harness it and move it in new directions. These divergent practices throw into comic relief, not the atrocities of trauma, but the commodification of that trauma into feel-able, bite-sized anecdotes of suffering designed to be absorbed between a hotel breakfast and an afternoon jug of beer. *Dancing Auschwitz* makes us see that any affective re-gesture towards the limits of representation marks both the tourist's embodied encounter with Holocaust history and the Holocaust performative as contemporary currency. In acting – but in acting differently – the *Dancing Auschwitz* performers participate in the same reclaiming of history that concentration camp memorials produce but also dislodge those memorials from producing, through unspeakability, other ideological interests and assumptions. It also remediates its act of remembrance into cyberspace, such that, separated from the space to which it 'belongs' and from how it should be 'experienced', the counter-affective currency of historical trauma circulates around the globe.

As I explained in this book's introduction, for Sara Ahmed, 'affects pass between bodies, affecting bodily surfaces or even how bodies surface'.[87] Affects, then, are not so much beholden to particular bodies as they are the staging of the 'betweenness' of bodies. As Schneider makes clear, the 'touching' of time enabled by gesture is what enables selves to unfold. Gestures are always (almost) self-touching in this way. Affects, including trauma affects, arise when our bodies touch (a site, another body, an image, a text, oneself). The unspeakable then, as one such affect-effect, is indebted to the gestural and the tactile, just as it is produced by its material effects. It is in practices such as those in *Dancing Auschwitz* that one's own affective complicity in the stagings of unspeakability can be re-felt. This might be considered as the conjectural space of inhabiting the suspension of signification, as feeling oneself feeling, or as the moment in which affect becomes meta-affect. It makes clear how feelings *about* feeling might dislodge some of the cultural affiliations that meld embodied experience to the responsible knowing of historical trauma. If affect is passed from body to body, produced by body in relation to body, and in relation to site and to history, then trauma sites not only modulate flows of feeling but also reanimate the feelings held therein. The Holocaust affect, as I have argued, mediates bodies' gestural repetitions with the aims and misfires animated by such sites.

2
Becoming Other-wise: Remembering Intercorporeal Indigeneity *Down Under*

Feeling white certainties

Australia is a country that takes the problem of its 'others' very seriously. That is, in the global discourse of Australia as a nation, Australia performs its relationship to otherness in serious ways.[1] Perhaps even more specifically, it could be said that Australia performs itself as a nation that is deeply concerned with otherness as a problem, and it does this as a way of creating an image of its self-composure. In shoring up its borders, literally and metaphorically, through languages that claim meanings about its others (and that therefore bring those others into being, often as immaterial others whose bodies do not matter), Australia produces a global sense of its own self-certainty. Australia produces itself as a nation that is strong and that can stand alongside other nations who also *other their others*. In this way, Australia is at its core aspirational, unsettled with its Antipodean sense of distant, 'down-underness', working hard to push that to the side so that it might perform itself as one of the same.

In this chapter I investigate how Australia creates Indigenous otherness as a means to enforce its mythology of nationhood, a mythology that has been practically maintained by various historical practices of colonisation, immigration, assimilation and multiculturalism, and also carried by the material effects of those policy discourses. More specifically, I am interested in how the idea of Indigenous memory is intertwined with a national sense, and *sensing*, of the nation as a landscape of cultural memory. Here, I examine how the Indigenous subject is positioned as a figure of memory, one who performs a 'deep' knowledge of the ancient past and who, in doing so, erases the trauma of colonial history. My argument emphasises the affective dimensions of what Chris Healy has called the 'contact zone' of Indigenous and non-Indigenous relations in the making of national discourses of 'Aboriginality'. For Healy, the term Aboriginality marks how the 'cultural and textual construction of things "Aboriginal"' occurs as an intercultural activity.[2] As Stephen Muecke points out, the promise of a relational viewpoint of Indigenous and non-Indigenous Australia means attending to 'discursive formations – linguistic and non-linguistic practices, institutional relations' such that one might discover those *'places where one's discourse is only made possible by its relation to the Other.'*[3] I argue that the affective landscape created by and through the Indigenous figure functions to shore up aspects of Australian belonging by enacting what Sara Ahmed calls 'non-performativity' in the name of cultural memory. I further suggest that it is in moments that arrest, decompose or *un-perform* this kind of non-performativity that a different set of affective meanings, and memories about those meanings, reveal themselves.

In my discussion, I move from the affective repertoire of Holocaust recollection discussed in Chapter 1 to the ethno-touristic setting of Australia's Tjapukai Aboriginal Cultural Park. I offer an inversion of how the manipulation of what I identified in Chapter 1 as the Holocaust affect, and more broadly as the memory affect, works. Founded over 20 years ago, Tjapukai Aboriginal Cultural Park bills itself as Australia's largest Indigenous cultural park, and is listed in the *Guinness Book of Records* for producing the country's longest-running stage show. The park offers interactive day and night tours, state-of-the-art audiovisual technologies and professionally choreographed dance performances. In its reconstruction of the corroboree alongside its portrayal of the Dreamtime, I argue that affect is used to evade and replace the traumas of colonial Australia with a popularised form of ancient cultural memory. Furthermore, I argue that at Tjapukai 'Aboriginality' is performed as sign and object of a national memory culture which positions the Indigenous memory of the deep past as a metonym for *the job of the nation remembering itself*. In this, the remembering Indigenous figure is scripted into a national memory repertoire, which sees Indigenous subjectivity as the symbol of a continuous act of timeless recall. I further suggest that this image of deep memory is *non-performative* in the sense that Sara Ahmed offers, enacting 'the failure of the speech act to do what it says [which] is not a failure of intent or even circumstance, but is actually what the speech act is doing'. Specifically:

> Such speech acts work as if they bring about what they name. Or, to be more precise, such speech acts are taken up as if they are performatives (as if they have brought about the effects that they name), which has its own effects.[4]

Via the idea of non-performativity, this chapter builds on observations of the rhetorical role of affect charted in Chapter 1 to position Tjapukai as a site that is experientially reconstructive of forms of nationalist cultural memory. I argue that the acts of memory that take place at Tjapukai replace historical recollection while performing as if they are genuine acts of recall. The dominant imagery of Indigenous memory mobilised at this site functions in such a way as to whitewash Australian history. These non-performative dimensions of the affective redeployment of Indigenous memory connect to what Elizabeth A. Povinelli has called a liberal politics of multiculturalism. In Povinelli's view, Australian multiculturalism operates as 'an ideology and practice of governance, a form of everyday affective association and identification and a specific discursive incitement across the variegated contexts of national and transnational life'.[5] For some scholars in the field, the multicultural frame does not apply to Indigenous discourse in the same way that it does to discourses of migrant ethnicity in Australia.[6] It is, however, within the particular logic of 'affective identification' that we can find that *both* are productive of feelings of what I call here *white certainty*, as Povinelli explains:

> As the nation stretches out its hands to ancient Aboriginal laws... Indigenous subjects are called on to perform an authentic difference in exchange for the good feelings of the nation and the reparative legislation of the state.[7]

In Kelly Jean Butler's terms, this exchange of 'good feelings' might be understood as a 'form of nation-building, [which] positions Indigenous peoples... as objects rather than subjects of feeling.

As such, it precludes their participation in the performance of ethical citizenship.'[8] In her work on witnessing, Butler observes 'a public mode of memory work', which has 'reshaped notions about what it means to be a "good" Australian in the twenty-first century'.[9] Importantly, it is in its ties to new forms of 'national imaginary' that this form of memory work 'too often works to affirm the position of settler Australians and solidify their affective ties to foundational myths of Australianness, such as the fair go'.[10] Like Butler, I argue that the popular rendition of the remembering Indigenous figure, one who is centrally, metonymically constructed as a landscape of cultural memory, is endowed with the affects that inversely produce feelings of being, and seeming, *white*. Here, the memory affect evokes the sensory certainties that contribute to shored-up forms of non-Indigenous selfhood under the pretence of engaging relational forms of exchange between Indigenous and non-Indigenous people.

In Chapter 1 I explored how the memory affect, working as the Holocaust affect, was used to transmit, commodify or even amplify traumatic memory; here I argue that the Tjapukai site conversely uses it to exclude trauma from the performative repertoire that enfolds it. When we understand the memory affect as that which *uses and stages* the force of feeling to perform transmittable acts of recall, it highlights how feeling is both medium and message of the meanings that Tjapukai mobilises. Chapter 1 described the memory affect as that which generates a specific kind of *feel-ability* around victims, and further, as that which describes *the felt effect of the truth-effect* of meanings about those victims. At Tjapukai, the memory affect accounts for how the nation's past is mobilised as a site of 'Dreaming', evocative of an experiential paradigm and property of Indigenous selfhood in pure terms.

As I will explain, the force of feeling contained in perceptions of an Indigenous 'sensing' of the past is used to confirm spectator subjectivities that produce sensibilities around Indigenous otherness. I therefore look at the ways in which tourists at Tjapukai are invited to 'feel' histories of Indigenous Australia via strategies of embodied induction. This kind of non-performativity works through a deeply historicised scenario of the encounter between Indigenous and non-Indigenous subjects that *acts over* the affective plurality such an exchange might otherwise evoke. Histories of colonial Australia and histories of Indigenous performance here reveal their genealogical interweaving, as evidenced in the park's own promise of servility, which states: 'We've been rehearsing for 40,000 years in order to showcase our culture to you.' In locating the production of felt history as the source of the site's commercial success, an account of *feeling* and *un-feeling* the memory of Aboriginality becomes key to sensing the relational ground through which cultural states of being, seeming and feeling *white* are made certain and unravel.

Sensing wise landscapes

Tjapukai Cultural Park is a short drive from the city of Cairns in Australia's tropical North Queensland, a tourist destination whose economy reaps the rewards of the famous UNESCO World Heritage-listed Daintree Rainforest and Great Barrier Reef. The broader landscape in which Tjapukai Park is situated is flanked by what Australia boasts as two of its greatest wonders – the 'pristine' waters of the Coral Sea, leading out to the Pacific Ocean, and the rich, fertile humidity of dense, rare forest. Tourists travel to Cairns for the balmy weather and fine food, for its fecund, exotic landscape and for the pleasure of a postcard holiday. It is in this context that the Tjapukai Dancers perform, and in which their performance in turn performs Indigeneity as a form of affective social memory.

Almost 15 years ago, Australia promoted itself as the globally spectacular host of the 2000 Olympic Games by calling upon the imagery of these natural wonders. As Deborah Rowe and John Stephenson note, the 'heavily charged atmosphere of the millennial Games' was an opportunity for 'a postcolonial, Southern Hemisphere nation about to celebrate its centenary, yet still struggling to reconcile with its Indigenous peoples... [to] apprais[e] and asses[s] contemporary "Australianness".'[11] Even above Australia's politics, the millennial Games were set to mark any host nation's success as a post-industrial cultural and economic milieu. While China's 2008 Opening Ceremony seemed to meet this expectation with its synchronous LED-lit Fou drums, Australia's imagery was less focussed on the splendour of technology and more on a natural world, one which it positioned as the site of its Indigenous culture, which it then positioned as the site of its national memory. In its Opening Ceremony, a young, white Hero Girl falls asleep to dream of a pre-colonial Australia. Gigantic oceanic puppets swim across the stadium, which swiftly transforms to become red desert, as Hero Girl awakens to 'an undeniable call from an ancient heritage, the Dreamtime spirits of another age' who incorporate 'over 40,000 years of culture, from [the country's] 600 Indigenous nations'.[12] Hero Girl is hustled across an earthen landscape by an Indigenous mob, to be met by an Elder who kneels beside her and gestures, evoking the skies. The sequence then closes as 'Nature' returns a landscape of flora and fauna with a burst of fire, evaporating the Indigenous performers from the scene. In this chronology, Aboriginal people appear in the nation's past, but not in its immediate future.

The Olympics Opening Ceremony was criticised for serving, as Michael Cohen *et al.* maintain, 'the interests of dominant power structures in its representation of minority cultures (Indigenous and migratory) in relation to mainstream Anglo-Celtic Australia'.[13] Specifically for Cohen *et al.*, it featured 'Aboriginality... through the eyes of an "angelic blonde"' which 'attempted to "white out" blemishes in Australia's "national script" and to discreetly "forget" areas of social sensitivity'.[14] It was, as they explain, 'a whitefellas' appropriation of blackfella dreaming',[15] or as Stevenson and Rowe suggest via Daryle Rigney, 'a way of creating a white alibi through "the construction of invader Dreaming – particularly [through] the perceived benefits of sport in healing the wounds of Indigenous people created by ideological policies of segregation, 'protection' and assimilation"'.[16] It is alongside the spectacle of the Dreamtime that the spectacle of sport became, as Catriona Elder *et al.* point out, '*the* space where reconciliation could and should take place'.[17] The success of the 400-metre women's final, in which Indigenous athlete Cathy Freeman performed as another kind of 'Hero Girl' to win gold and stop the nation, was viewed 'as a moment of profound political significance'. The 'Freeman Final' was one such *non-performative* that promised to, as Freeman's NIKE slogan had also anticipated, '[c]hange the world 400 metres at a time'.[18]

The Opening Ceremony and the Freeman Final can be understood as examples of what Diana Taylor calls 'scenarios', in which cultural repertoires in the present reveal their entanglement with a preceding dynamics of colonial power. As Taylor explains, scenarios offer a framework for understanding how a dynamics of Indigenous/non-Indigenous relations is not only discursively, but also affectively, transmitted across time. In emphasising gestural repetition alongside textual storying, Taylor suggests that scenarios account for the transfer of cultural memory through embodied practices as well as the containment of the embodied by frameworks that recall, and emphasise, archival logics of memory. Specifically, the scenario is that which 'includes features well theorized in literary analysis, such as narrative and plot' but requires that 'we also pay attention to milieux and corporeal behaviours such as gestures, attitudes, and tones not reducible to language'. Scenarios hence 'frame and activate social dramas' and are 'the product of economic, political, and social structures that they, in turn, tend to reproduce'.[19] In this sense, they point to 'the social construction of bodies in particular contexts'[20] and pose a way to understand the transmission of meaning across those bodies and how those bodies are discursively positioned as beholders of meaning. Scenarios can thus be *non-performative* in the sense that Sara Ahmed argues, working '*as if* they bring about what they name'.[21] For Taylor, the scenario of the colonial encounter is one that persists in staging 'the West's perspective – the novelist, the playwright, the discoverer, or the government official – it stars the same white male protagonist-subject and the same brown "found" object'.[22]

Taylor's emphasis on scenario over narrative makes clear that the 'brown "found" object' is materially produced as well as being productive of certain material effects: particularly those effects characterised as producing feelings about being white. That is, 'finding' the object is itself a verb that requires lived actions in material places performed by real bodies. Finding the brown object involves a sensing of that object and, in turn, a sensing of oneself through the felt co-production of that object as having been 'found' or as being 'find-able'. In the Australian example, as Povinelli adds, the scenario of the colonial encounter which stages the find-able brown object entails less a requirement that 'colonized subjects' are discursively inspired 'to identity with their colonizers' and more a sense that

> Australian... multicultural domination seems to work... by inspiring subaltern and minority subjects to identify with the impossible object of an authentic self-identity; in the case of Indigenous Australians, a domesticated nonconflictual 'traditional' form of sociality and (inter)subjectivity.[23]

Here, we understand that the evocation of the affective space of ancient tradition used in scenarios of encounter, which displays white wonderment at euphemisms of 'found brown' knowledge, is a socially practiced mythology recalling an ancient past to the centre of national memory. Indigenous Australians are positioned to assume possession of this ancient past, *to be cultural rememberers* in place of the white nation remembering itself. Indeed, it might be said that the white nation *delegates* the act of remembering to its Indigenous people at precisely the points where the terms of its historical reflection become uncomfortable.

In this sense, the non-performativity of the kinds of cultural memory that are produced around the figure of the Indigenous rememberer relies upon that figure producing itself as a form of affective certainty. This is a kind of certainty that Povinelli additionally notes is established as an impossible ideal against which any Indigenous subject might fail. Instead, as Povinelli explains, the imagery of Indigenous tradition in contemporary Australia travels in scenarios that enact a specific form of state machinery, which demands that

> [t]o be truly Aboriginal, Indigenous persons must not only occupy a place in a semiotically determined social space, they must also identify with, desire to communicate (convey in words, practices, and feelings), and, to some satisfactory degree, lament the loss of the ancient customs that define(d) their difference.[24]

The imagery of Indigenous tradition hence operates to engineer those effects that are produced *in the name of memory* but that are not themselves productive of memory. It is, as I made clear in this book's introduction, demonstrative of a kind of memory that can be understood as a memory affect: a feeling of memory that operates in the name of memory, that generates a feeling of the occurrence of memory, but that negates memory in doing so. In this instance, it imagines 'the ancient traditions of Aboriginal people as a powerful, pleasurable, persisting force predating the nation and defining its historically specific difference'. Here, Indigenous people are 'not only distinguished by their genealogical relation to the nation-state but also by their affective, ideational, and practical attachment to their prior customs'.[25] Indigeneity *stands in* for cultural memory in a way that enables the nation to erase the history that erased traditional practices, or confined them to the repertoire.

This is to argue that the memory of Australia's Indigeneity and, doubly, the memory enabled by the remembering Indigenous figure are endowed with a unique phenomenal horizon given in assumptions that Aboriginal people sense landscape and time in different ways.[26] The evocation of 'Indigenous tradition' therefore carries with it an affective space of *union* with the landscape that poses invitations to *feel* its *otherness* in ways that promise to open out a new kind of sensory terrain. This form of cultural memory calls to the beginning of time, it is 'supra-' temporal in that sense. It extends beyond living memory to encompass myth as marking a time that existed before, or beyond, itself; it offers a promise of what it is to 'experience' time on new terms. Interestingly, Povinelli tracks this fascination with the Indigenous sensorium to an 1880 ethnology written by Reverend Lorimer Fison, who observed in the sexual practices of the Indigenous Kamilaroi and Kunai 'ancient rules' which entailed a sensing of 'something else,...something more.' Povinelli calls this lure to the 'something more' a 'some *thing* that offered him [Fison]...a glimpse of an ancient order puncturing the present, often hybrid and degenerate, Indigenous social horizon'.[27] She hence argues that held in the symbology of Indigenous tradition is the assertion that 'Aboriginal subjects should...construct a sensorium in which the rest of the nation can experience...a national noumenal fantasy'.[28] A sensing of a sensation *that is sensation on new terms* – that calls to a sensory paradigm that enacts in the body a new *sense of itself* – is the promise put forward by calls to traditional Indigenous culture made by the nation's dominant memory repertoire. I suggest that this form of affective certainty is enacted through logics that recall so-called ancient memory to replace historical recollection.

If the performing Indigenous subject promises an explosion of the 'noumenal' *something other* or *else* in its non-Indigenous viewers, then this is a dynamic that the scenario of encounter reinscribes through discursive as well as affective dramaturgical logics. The scenario of encounter not only aims to envision this something other or else, but it aims to restage it as a *dynamic of contagion* produced by the encounter as a social field of recall. The dominant imagery around this noumenality might be seen to be archetypically conveyed in scenes such as those in the Opening Ceremony, in which we see Hero

Girl infected and affected by her encounter with Indigenous bodies, which themselves evoke 'mysterious' meanings of the 'found brown' land that she inhabits. In the analysis that follows, I suggest that this imagery of encounter, as well as its aim to implant a prosthetic kind of ancient memory in its non-Indigenous interlocutor, is made experiential and participatory in a site such as Tjapukai Aboriginal Cultural Park. The promise of a park such as Tjapukai is that visitors are invited to feel, momentarily, like Hero Girl, who momentarily feels a version of her own white fantasy.

Performing Indigeneity

There are two compelling histories that depict relationships between the dynamics of settler Australia, traditions of performed Indigeneity and the popular production of an Indigenous 'noumenal' fantasy. These histories provide varying inflections on how the scenario of encounter evolved as an effect, as well as an account of Indigenous and non-Indigenous relations. In the first, Roslyn Poignant charts the practices of ethnographic display that took place with the rise of the modern spectacle. Poignant unearths the forgotten histories of two groups of Northern Queensland Aborigines who were abducted by the theatrical agent Robert A. Cunningham in 1883 and 1892 and sold to the American showman Phineas T. Barnum to tour to the metropolitan centres of America and Europe for display at the World's Fairs.[29] Both groups (the first including nine Aborigines, the second eight – six men and two women) were from separate communities on Palm and Hinchinbrook Islands – they did not all speak the same traditional languages, and not many spoke English. The first group performed at the *Ethnological Congress of Strange and Savage Tribes*, forming part of Barnum & Bailey's *Greatest Show on Earth*, to then endure a humiliating season of dime museums before being sent to the Folies Bergère and the Société d'Anthropologie in Paris, where members Tambo and Wangong died. The second group toured to the Columbian World's Fair of 1893, and together both groups spanned 16 years on the touring circuit.

Poignant recounts that the Aborigines immediately became 'enmeshed in Western systems of mass entertainment and education...which marked the emergence of the modern world as spectacle'.[30] She begins her history with the contemporary repercussions of these practices – the 1994 repatriation of Tambo's bones, found in a funeral home in Cleveland, Ohio (where they had been formerly mummified and displayed in a museum). In tracing this final act of repatriation back to its performative contexts, which are spoken only by the unsettling ghosts of anthropological photographs, Poignant describes how the performers' 'troubled expressions' sit uncomfortably against the 'backdrop of a European pastoral scene', displaying 'body language... [which] conveys an air of resistance to the proceedings.'[31] In short, she sees the practices of spectatorship that colonising modernity produced, such that '[w]ithin their performance roles, [the Aborigines] were simultaneously themselves and reflections of the "savages" of western imagination'.[32] They appear as ' "likenesses" of particular people who were once alive and stood in front of the camera'.[33] In her comparable analysis, Jane R. Goodall finds that there were contradictory dramaturgies contained in the concoction of such 'liknesses', observing the measured balance a performer needed to strike between skilfully performing 'natural' Indigeneity, while suppressing the skilfulness of that task. Goodall points out that 'a successful portrayal of the primitive acting naturally involved a complex set of skills related to sophisticated forms of intentionality'. In short, 'a too-skilled presentation of savagery might paradoxically fail in its objectives'.[34] The cultural space of repatriation hence signals an unfolding of what Poignant calls the modern European 'show-space' which, like the scenario, operated as 'a zone of displacement for the performers and a zone of spectacle for the onlookers'.[35]

These accounts of skilled intentionality give weight to the histories of Indigenous performance spectacle that evolved in settler Australia alongside European modernity. What Michael Parsons calls the 'tourist corroboree' could be viewed as one such example of a performance repertoire which, from its inception, was a product of showmanship. In his study of its emergence in South Australia and the Victorian town of Ballarat, Parsons finds that Indigenous performers were well in control of the performance apparatus that contained them. That is, the corroboree operated as a 'public dance-music even[t] staged by Aboriginal people for a settler audience', and was *specifically* an 'Aboriginal-initiated and organised... attempt to use symbolic goods to engage in the settler economy'.[36] Parsons further explains that any viewpoint that reads the tourist corroboree as a

' "prostitution" of the "real" ' fails to understand its validity 'as a new cultural artefact in its own right'.[37] It became, as he explains, a significant tourist attraction throughout the 20th century and 'a major source of commenting upon and critiquing the new settler society'.[38]

In view of these accounts, we understand that the socio-historical emergence of the show-space of Aboriginality describes not only histories of agency and oppression, but a symbology of what Stephen Muecke has since called the contemporary burden of 'a totalising concept of Aboriginal culture', which operates as follows:

> Normally, when one lives happily in a culture one 'swims' in it; one is not constantly reminded, as they are, that one is (Aboriginally) cultural. This legacy forces contemporary Aboriginal subjects, in turn, into positions of essentialism (you *are* Aboriginal), or representativeness and knowledge (*you* would know about kinship systems of the Western Desert), and consequently they are constantly called upon to display this essence, or this or that skill, as if culture were an endowment.[39]

The ancient memory that a park such as Tjpaukai invokes can be seen to have been built from this history, a history which the park neglects to fully encompass, but which nonetheless forces itself onto the show-space to disrupt the affective logics of forgetting that are at play. In what follows I suggest that Tjapukai Cultural Park brings to the present the historical dynamics contained in practices of performed Indigeneity, which simultaneously evoke histories of skilled intentionality as well as histories of commercial strategism, both catering to a specific neo-colonialist eye. What spills onto the affective show-space of the park is a subtly unconscious enactment of that very duality.

Framing Aboriginal ethno-tourism

Existing studies on Tjapukai Aboriginal Cultural Park examine the questionable and conflicting impacts the park has had on its local community. In these studies, the framework of ethno-tourism is either understood for the commercial benefits it brings to the local Djapugay people or for the cultural value it provides for tourists, often at the expense of traditional Djapugay culture. One study characterised the rise of the ethno-tourist model in Australia as a relatively recent, but rapidly expanding, industry, with an estimated value exceeding 100 million dollars per year.[40] As it explains, the sector caters to vastly different market desires, which

> [n]o longer just about dance and didgeridoos... now cover almost every facet of contemporary tourism: cultural, eco, adventure, volunteering, luxury stays, camping. You can rescue green sea turtles on Cape York Peninsula, follow songlines in the outback, visit the cellar door of an Indigenous winery beside the Lachlan River in NSW, take a cultural cruise on Sydney Harbour, stay in a safari tent on the fringe of a Central Australian community, or participate in a traditional hunt.[41]

Ethno-tourism tends to be most successful when offering boutique packages that allow for 'intimate and authentic experiences',[42] which attract the high end of the market, for example, the safari tents priced at over 1,000 dollars a night at the edge of the Titjikala community, or the Sounds of Silence starlit desert dinner offered at Yulara.[43] Interestingly, the bulk of the ethno-tourist market is international, with the suggestion that overseas visitors are less interested in the educational aspects of cultural tourism, and more in how the symbolic capital of Indigeneity can purchase an experience of the 'outback':

> Attractions based upon Aboriginal culture may not be as popular as is generally thought,... their presence helps an evocation of the outback, of which Aboriginal culture is a part.[44]

Tjapukai Aboriginal Cultural Park is one of the nation's founding capital ventures into this very complex marketplace. It might in this way be characterised as a comparatively 'slick tourist production' in the face of other, more nuanced and less 'packaged' ethno-tourist experiences.[45] That said, the park celebrates its history as Australia's longest-running stage show, owing its 1938 naming, Tjapukai, to anthropologist Norman Tindale[46] and its evolution to its beginnings as the nation's first ethno-tourism experiment, as its website explains:

> In 1987, Don and Judy Freeman, international theatre artists along with partners David and Cindy Hudson and 5 local

Tjapukai men; Willie Brim, Alby Baird, Wayne Nicols, Irwin Riley, Neville Hobbler and Dion Riley created the first Aboriginal dance theatre.... In 1996... Tjapukai was relocated to its current site at Caravonica, and expanded its business activities to include interactive demonstrations and performances, a cultural village, restaurants, retail and much more.[47]

Having evolved from the historically successful Tjapukai Dance Theatre originally located in Kuranda, Tjapukai Park now covers 25 acres near Cairns. It builds on this heritage to bring benefits to the local Djapugay people, with the park website stating it has 'injected in excess of $35 million to the local Aboriginal community in wages, royalties and through the purchase and commissioning of art and artifacts' as well as ensuring that the benefits of cultural tourism are conferred directly back to the people by making Aboriginal tribal councils majority shareholders.[48]

Studies on the success of Tjapukai Park as an ethno-tourist model compare the satisfaction of tourists with the effects on the local community. These effects include, as Dyer *et al.* point out, discrepancies in the way the equity of the park is distributed, misperceptions around the involvement of Djapugay in management roles and the perceived lack of involvement of Djapugay in curatorial decisions. In short, while the park has promised that it maintains a 51 per cent Djapugay shareholder base, that it employs primarily Djapugay people and that it is the only park in the region to celebrate traditional Djapugay culture, it has also been accused of watering down culture by employing non-Djapugay and by operating a shareholder structure that diffuses Djapugay equity too broadly. Such conflicts appear in accounts from employees that 'any Islander or Aborigine can work here'.[49] There are further accusations from employees that the park capitalises on the Djapugay name, while ignoring the history of the removal of the Djapugay from their homeland and onto missions. In this context, performers' concerns at inaccuracies in the representation of their culture, in which 'Park managers expected them to change their style of dance in ways that were contrary to that taught by their grandfathers'[50] seem to carry, rather than negate, the history of the show-space of Aboriginality as a colonial power structure. Perceived inaccuracies in the performance of culture include adjustments to traditional women's dances to suit tourist responses and the use of the didgeridoo, which was neither made nor played by traditional Djapugay. At the same time, employees value the park in continuing the presentation of Indigenous culture for both Indigenous and non-Indigenous communities.

Unbecoming scenarios and 'affect-accidents'

The 25 acres of Tjapukai Park are not arid desert landscape such as that featured in the Olympics Opening Ceremony, and they are not the tropical wetland that frames the park's immediate surrounds. They are a landscaped, serene parkland, joining outdoor activity areas for boomerang and spear throwing, and outdoor performance areas such as the Tjapukai dance theatre and activities hut, across a bridge and river, to indoor buildings, which house a creation theatre, a museum, a restaurant and a souvenir shop and gallery. Daytime tours travel a circuit around these various stages, performing traditional music and dance routines, bush tucker and bush medicine education sessions, a holographic representation of the Dreamtime story of the cassowary, and spear- and boomerang-throwing classes. The park also boasts a recent addition to its programme, a 'Tjapukai by Night' evening tour, which includes a corroboree, a fireworks ceremony, a dinner and a live band. I participated in both day and evening Tjapukai events. As I did so, I watched Tjapukai dancers perform, I learned about bush medicine and weapons as well as Indigenous law; I also practised throwing a spear and a boomerang. These forms of interaction were familiar repertoires of performed Indigeneity in terms of the accounts given in Poignant's and Parsons' studies. It was, however, at Tjapukai by Night that I felt the park animating this repertoire through a form of spectacle that accidentally revealed its non-performativity. That is, it performed a scenario that *un-became* itself: that revealed an altogether different series of stage affects. For this reason, it is the Tjapukai by Night tour that I will focus on in the remainder of this chapter, and I begin by turning to the language of feeling evoked by the park's website:

> Feel the powerful spirit of Indigenous Australia vibrate through your body as you embark on a mysterious journey into the ancient Tjapukai culture by night...
>
> The Tjapukai Aboriginal performers will lead you into the Magic Space for an emotive didgeridoo blessing ceremony and the

mystical Quinkan spirit performance. Your initiation continues with traditional face painting as the Tjapukai performers escort you to the lake...

With clapsticks in hand you add to the pulsating rhythm of the corroboree while Tjapukai Aboriginal warriors use traditional fire sticks to create fire. Suddenly a spear is launched and a spectacular fire ball explodes into the night sky.[51]

Affect is everywhere at Tjapukai. With its high rhetoric of the Indigenous Dreamtime as that which spectators can 'feel', 'journey into', 'pulsate to', Tjapukai promises a dramaturgy that recalls the Aboriginal show-space as the vehicle of Fison's much-desired noumenal fantasy, the 'something else,... something more'. In doing so, it mobilises a series of affects that are designed to communicate the feelability of 'found brown' bodies and their socio-specific knowledge. I suggest, however, that this is a dramaturgy that draws upon known repertoire to erase real memory effects. In being non-performative, these 'pulsating' affects insert a scenario of timelessness into a place that might otherwise engage a historically *emplaced* account of cultural remembering. In this sense, the scenario of encounter that Tjapukai stages is constructed as a form of affective contagion that both transmits, and erases, a memory of Australia's colonial history. In particular, it omits to include the very memory that knows that what we see is a production of, as Taylor suggests, a deeply historicised dynamics of power. However, on the evening that I saw Tjapukai by Night, known repertoire failed, and when it did there emerged a glimpse, or 'affective spill', of memory's *affect*. By this I mean that the performativities of memory as it was staged as a process that engages, constitutes and produces the *felt* certainties of a determining sociality, emerged as a sensorial account of *affect's affective dimensions*. This *sensing* of the failure of the spectacle of ancient Indigeneity – indeed, the moment that could be called the undoing of the 'noumenal fantasy' – became flinchingly, if accidentally, apparent, and was the park's most powerful and positive, affective social force.

In focussing on ideas of upset in relation to spectatorial affect, I draw on Nicholas Ridout's account of theatrical accidents, which characterises that moment when the machinic components of a stage drama, here understood as the product of a social drama, break down. Ridout argues that accidents have the capacity to reveal theatre's essential 'ontological queasiness' by being moments 'which seem both to underpin and undermine the functioning of theatre as a mode of ethical or political communication'.[52] In theorising affect as that which the moment of theatrical collapse activates, Ridout points to theatre as a 'machine' of 'labour' whose undoing can breed 'a politics rooted in shame rather than mutuality' and moreover, 'an apprehension of our own position in relation to the economic and political conditions of our theatre-going'.[53] The kinds of machinic labour that underpin the show-space of Aboriginality can be seen in how the Indigenous performer is given to behold a cultural endowment that speaks to a 'powerful, pleasurable, persisting force... distinguished by... practical attachment to their prior customs'.[54] The dramaturgy that promises the affective contagion of found brown bodies, then, is one that produces their *feel-ability* in relation to white, feeling, sensing bodies. In the case of Tjapukai, the accident opens out a spectatorial experience of feeling, not otherness, but *other-wise*. This, I suggest, is the affective spectre of the collapse of the sensorial certainties that amount to feeling and being white. Feeling *other-wise* denotes the perceptual reckoning with one's own subjectivity that occurs when the framework of a sentimental sensorium falls away. It is when one feels neither white nor brown, and so may come to an awakening of their own subjective uncertainty.

Playing with fire

The Tjapukai evening begins with a glass of wine in the art and souvenir shop. Soon we are moved into the museum, which houses a number of traditional artworks that are discussed by guides. The park's mission statement is also proudly displayed on a plaque that reads:

> The Djabuganydji communities and their elders have approved and overseen the material presented in the park. They own the land the park is situated on and have a substantial equity interest in the venture.
>
> There is no other presentation of Aboriginal Culture in the area which provides any benefit to our community or which has

requested or received the authority to present our culture for profit. Tjapukai Cultural Park is the only authorized representation of Aboriginal Culture in the Tjapukai Tribal area.

The first such 'authorized representation' appears when the lighting dims. A booming voice, set against sounds of cracking thunder, narrates: 'I am Gadja, the spirit of the land – Djapugay land. I am the earth, the sun, the rocks, the trees. I am in the rivers. I am in the mountains.' Voice number two then appears: 'I am Quinkan. I am the rain and the floods. I am the storm and the thunder. Hear my voice. Don't mess with me.' In darkness, two figures, the first, Gadja, animated by a dancer dressed in blue and yellow fluorescent luminescent frills, and the second, Quinkan, an inflated puppet stick-like figure, larger than life, appear in the darkened space. They narrate a story of origins through shadows and smoke, which eventually give way to the fading outdoor sunset, as we are walked outside and given clapping sticks and face paint to participate in a corroboree.

'G'day,' we are told by performers standing on a small earth platform that we circle, 'all Djapugay ceremonies, or corroborees, began by making a fire.' The weather has been damp and raining. Our clapping and singing will provide momentum for the performers' display of traditional skill. They teach us a song, while taking turns in a ceremony that uses traditional fire sticks to create fire. The sticks swivel and the kindling smokes and coughs. We sing and clap. The kindling smokes and coughs but fails to light as the performers keep smoking and blowing and smoking and blowing, our attention turned, eventually, away from the imminent fire and towards ourselves as tourists, from many backgrounds it would seem, who bring along toddlers, teenagers and grandparents, and who frame the traditional dress of the male performers with converse contraptions of video cameras, travel bags and variable expressions of enthusiasm or discomfort with our enforced participation. Watching the sticks swivel, it becomes clear that the performers are becoming tired. They persist. In my clapping, I feel the blisters that must be appearing on their hands with the rubbing and turning of the fire sticks. With the absence of fire, momentum wanes, and with a breath of comic timing, a performer calls out to us all in genuine appeal: 'Anyone got a lighter?' There is a pause, but the invitation is met with silence. It seems no one quite knows how to respond. The performers return to their swivelling, the audience to their limping clapping, until finally a guide exits to return with a package of already lit kindling that sends the scene, thankfully ignited, on its way.

In his history of the tourist corroboree, Parsons finds one early description of performing with fire in an account given by the New South Wales Surveyor-General from 1828–55, Thomas Mitchell, which evokes the thrall in which such performances were held:

> The dance always takes place at night, by the lights of the blazing boughs, and to time beaten on stretched skins, accompanied by a song. The dancers paint themselves white... Darkness seems essential to the effect of the whole; and the painted figures coming forward in mystic order from the obscurity of the background... have a highly theatrical effect.[55]

As Parsons explains, the social and economic milieu of settler Australia framed Aboriginal performance through Western theatrical experience, 'which at that time included a great emphasis on spectacular effects'.[56] In fact, Parsons notes that, by 1847, payment for Sunday corroborees was an established cultural practice, citing Protector of Aborigines Matthew Moorhouse's submission that his warnings to cease the ceremonies on grounds of religious observance were not being heeded.[57] Indeed, it is in the example of a 1867 Ballarat corroboree that incorporated the use of pyrotechnic displays, and which created 'a curious effect... produced by the glare of the colored lights thrown upon the group as they went through their diabolic looking manoeuvres',[58] that the *scenario* of the corroboree as tourist performance becomes apparent.

No sooner does the kindling give spark to the burst of fire on the Tjapukai performance podium, than the fire travels into the air, carried on a burning spear that lands in the black humid distance of the park with an almighty boom. The spear has been sent towards an ammunition store, which erupts in a spectacular fireball that momentarily illuminates the still nightscape. The flash is unexpected, a blast of grandeur equating to the luminous brilliance of other, more familiar Western fireworks spectacles. For Maryrose Casey, fire in Indigenous ritual and symbology connects land and people. It has been used in contemporary times to draw the past into the present as an act of resistance, such that '[r]itual fires and associated ceremonies...have been a common feature of Indigenous actions over the last decade'.[59] It is in this way, as Marcia Langton explains, amongst water and earth as one 'sacred and elemental source and symbol of life', an element that is 'meaningfully entangled in social relations'.[60] Tjapukai's unexpected spear-lit fire shoots this very entanglement across the landscape, enveloping spectators and performers in a splay of illumination that seems to want to punctuate all bodies and material elements with each other, producing a sensing of ourselves as one, even as 'atmosphere'. With a round of applause, and the fireworks successfully aglow, we are ushered into the dining room where an international buffet awaits us, and where the Tjapukai performers continue to sing and dance on stage.

Given Parsons' observations on the use of fireworks as evidence of a highly crafted symbology of Indigeneity and of a precisely modern mercantile dramaturgy, we might look to the kinds of affects that this failed traditional, then redeemed pyrotechnic, spectacle mobilises, and hence, to the kinds of affects-about-those-affects which appear when the fire fails to burn. As an act that recalls the tourist corroboree to contemporary life, the Tjapukai fire echoes those early negotiations that forged the Aboriginal show-space of settler Australia. Its pyrotechnic display can, in particular, be read as an echo of events such as the 1867 Ballarat corroboree described above. Importantly, the 1867 use of pyrotechnics might be understood for how its novelty produced, *en masse*, a demonstration of brown man's mastery over white man's fire-power, which as Simon Wherrett notes, was itself a demonstration of white man's '*power over fire*' whose elemental 'divine and magical connotations were strong' in (post-) Enlightenment modernity. In Wherrett's study, the cultural evolution of pyrotechnics reveals that, as spectacular experiments with a scientifically contained natural force, fireworks made visible the desire to rationalise the supra-natural, such that explosive forces aimed 'to bring the heavens down to earth' in a display of authority.[61]

The noumenal fantasy evoked by the use of fireworks at Tjapukai explodes the strange doubleness of these traces: it mobilises fire as a container for those affects which 'belong' to Indigenous bodies and at the same time travels modernity's 'scientific' capture of those very inflammatory characteristics of so-called irrational others. The Tjapukai fireworks here evoke a performance genealogy that constructs the illuminating desire of white man to at once spectacularise and cleanse the darkly Indigenous bodily sensorium in a moment of transmitted, *ignited*, feeling.

This use of fire as a symbol that mobilises those affective dimensions of an ultimate Indigenous Otherness materialises what Patrick Wolfe, in his anthropology of the anthropology of the Dreamtime, calls the 'Dreaming complex'. For Wolfe, the Dreaming Complex considers 'an anthropological construct called the Dreamtime, and not any presumptive Koori precedent',[62] pointing out that in its combination of trans-temporality and psychic life, it emerged at the same time as Darwinian notions of natural selection and psychoanalytic notions of the unconscious. In doing so, it enfolded Aboriginal people in a timelessness that constructs ideals of unison between self and world. Specifically, Dreaming as a concept both disregards 'sequential regularity [running] counter to the whole [modern] discourse of time, discipline and order' and also represents the 'abstract...notion of a soul, or of a spiritual double detachable from the body, which occurred to savages as an explanation for the sensation of moving about in their dreams'.[63] In this way, as Wolfe explains, this 'unquestioned proximity to the animal state' presumed that Aborigines 'like animals, should be held to confuse dreaming with everyday experience'.[64] They became, in Wolfe's words: 'hapless dupes of somatic caprice' who solely 'thought with their bodies'.[65]

The semiotics of the Dreamtime for Wolfe signify an ultimately somatic form of intelligence: a unison with atmosphere; a fire exploding in the darkened night. As a version of white man's perceptions of found brown bodies, the material implications of this 'theme of precolonial somnambulance'[66] can be seen in how it has worked to solidify arguments for *terra nullius* and hence for the inception of settler societies to bring the land to consciousness through *timely* white bodies who *keep in time*.

Remembering Indigenous remembering

Perhaps the cultural evocation of Indigenous people as being *without time* is a way to position Indigenous memory without a history. That is to say, the evocation of the Dreaming complex positions the past as a sensed, or sensory mechanism without positioning that evocation as itself a product of historical specificity. In this argument, the memory affect can be held to account for mobilising memory in the place of history; for inverting the parameters by which the more usual opposition of small 'm' memory and its micro-narratives in relation to capital 'H' History and its authorising forces are perceived to operate. The memory affect here evaporates history from the scene of a staged sensorium, which, like the Olympics Opening Ceremony, positions Indigenous people as pre- and post-chronology. This evocation of an intense sensibility towards a kind of ultimate timelessness disables Indigenous people from moving beyond the affective logics of a show-space that holds them as always already feeling beings. One might further surmise that this very performance of Indigenous bodies assumes an ultimate affective potentiality in which they discursively demonstrate the pre-personal knowingness of affect itself.

Arguing that the affective momentum staged at a site such as Tjapukai has been fabricated for the purposes of whitewashing history is, however, not to demand that a viewpoint of what Gillian Cowlishaw has called the 'nation's favourite wounded subjects'[67] be instead brought to the foreground. One only need turn to the varieties of public sentiment that surrounded the historic 'Apology to the Stolen Generations' to understand the vulnerability and scepticism with which narratives of Indigenous memory are met. As Cowlishaw notes, this discourse itself risks creating another set of wound-affects, in which '[a]ttributing a common history of pain and suffering to Aboriginal people...positions them as inherently needy and damaged in some abstract and disembodied way that is...dehumanising'.[68]

The fire display at Tjapukai aimed to make us all awash with a feeling of timelessness evoked by the puncturing of the dark black night landscape with bursts of fire, sight and light. At the same time, its evocation of ideas of phenomenal immersion revealed the failure of affect to take charge. Indeed, in the less than pulsating fire ceremony on the night that I was watching, the emergence of the unfamiliar, but clearly backstage figure of the contemporary Indigenous *bloke* who wants *a lighter* busted open any form of machinic certainty the stage/social drama might have wanted to keep intact. The silence at his call for a lighter revealed the spectre of our mutual entanglements in a drama that no one seemed to feel much like playing at. Feeling *other-wise* to the moment of performance here generated affective plurality and perhaps even surprise at the spectre of the rather ordinary, less than pyrotechnic, less than fluorescent, man before us. In the moment of his speech, we felt, as Muecke intimated, the place *'where one's discourse is only made possible by its relation to the Other'*. A dramaturgy that stages this place of potentiality more often might accommodate both discourses around phenomenal certainties and the certainties of those discourses in other performative ways. Indeed, a dramaturgy which points to the decided historical fact of the corroboree's atemporality could reframe the way in which we experience the memory affect as a specific assemblage of 'whitening' tourist time. This form of dramaturgy would harness the accidental affects that percolate at the edges of this tired, and tightly bursting, scripted repertoire. It would stage a continuing affective state of becoming *other-wise,* which could enable a socio-cultural dynamics of memory more hopefully operative in a future Australia.

3
Feeling Remediated: The Emotional Afterlife of Psychic Trauma TV

Takeaway trauma

Sydney's Entertainment Centre holds almost 13,000 people. Add to those people their dead relatives. Add again a few coincidental connections to the dead friends of those dead relatives, and combining the dead and the alive, psychic medium John Edward has an enormous crowd to please. In my youth, I spent some years watching Edward's television programme *Crossing Over with John Edward* on cable TV. But now Edward is here in person. His website claims that this Sydney tour sold out in 15 minutes, placing him amongst the vendor's bestselling acts of all time. A murmur erupts as the stage compère arrives to 'rev us up'. We have come to see John Edward in action. *Clap.* We have come to witness the *real thing. Clap.* We have come to be in the presence of dead people. *Clap.* We have even come to talk to a few of our own. *Clap... clap... clap...* and with that, John Edward – the world's first self-proclaimed 'media medium' – enters the arena.[1]

In this chapter I move from the site-specific tourist productions of the memory affect described in chapters 1 and 2, to the televisually constructed landscape of generic grief performed in the United States-based, popular psychic television programme *Crossing Over with John Edward*. As a dually live and televisual event, *Crossing Over* is a contemporary rendition of a traditionally arcane practice. Psychics have the job of performing that which nobody else can master: making the disappeared reappear. In the West, this most commonly happens in the occult fairs and carnival sideshows that house the taboo.

I suggest here that *Crossing Over* repositions the practice of ghost mediumship – a performance lineage most often aligned with premodern mystic arts – alongside the differently spectral modalities of technological postmodernity. In doing so, Edward's ghosts perform an affective reckoning with present concerns. They arrive under the guise of enabling mourning to enact a precisely staged apparatus for maintaining normative social selves: they *appear* in order to *disappear* the mechanics of affect that the programme relies upon for spectacular effect. Edward has joked that *Crossing Over* resembles a version of 'Opie does Dead People' – a curious blend of reality TV vernacular, game and talk show genres with paranormal content to boot.[2] Tabloid writers have dubbed the programme 'a monstrous hybrid of talk show and psychic fair, spliced with the look and sound of a 900-line infomercial',[3] pitting Edward as a 'tomb reader', 'medium rare' and 'hustler' of the bereaved who exploits emotions for commercial gain. In this chapter I consider Edward's ghosts as neither real nor fake. As memory affects, they produce real effects, exposing the cultural enmeshment of felt, personal trauma with its spectacularised televisual afterlife.

This chapter builds on the popular cultures of Holocaust and ethno-tourist stagings of cultural memory mapped so far in this book to underscore the memory affect as an experience of the traverse across live and televisual spaces of feeling: spaces which are haunted in complex ways by the remediating reciprocities they pre-produce for each other. While in Chapter 2 the ancient scenography of the Australian 'Dreaming complex' was considered as a form of national forgetting, here I consider the spiritual postmodern as not so much past-orientated as precognitive. I argue that it mobilises late capitalist models of subjecthood that affirm personal trauma as a publicly performable form of sociality, while at the same time using it to promise a future without trauma. I also identify Edward's spectators as publics who *tune in* to their own construction of the memories of others as a means to validate normative ways of being. Mourning here becomes an enclosed affect-driven circuit of consumption and catharsis. In marking the John Edward experience as at once pre- and post-temporal – as a kind of felt potentiality of subjective certainties – I draw upon Richard Grusin's notion of the premediating logic of post-September 11 televisual time and space and connect this to the explicit theme of ghost mediumship that Edward strikes.

The memory affect here reveals that the *Crossing Over* ghosts, as a specifically contemporary form of memory culture, make trauma acutely political by constituting publics who collectivise as *bodies of sentiment* around an assumed sensorial afterlife. These are ghosts that materialise what is in Avery F. Gordon's terms 'a shared structure of feeling...a specific type of sociality'.[4] They are 'double-spooks', taking the form of 'that which appears absent' but which is rather a dangerously 'seething presence'.[5]

Below, I argue that *Crossing Over* offers a ghost world that produces communally contagious feelings of trauma as they come to be commercially shared and televisually sold. I further suggest that the memories mobilised by Edward's programme can be understood in their specificity as *predeterminations of feeling* that are designed to be felt as *corporeal after-effects*. In making this claim, I examine how Edward engages his live audience in the embodied co-production of a collectively narrated ghost world, a process that involves the textual creation of memory traces to be affectively interpreted by his studio spectators. His live gallery summons ghosts as a way of enabling audience participants to air individual fantasies of remembrance, to recollect histories of trauma, and by which they can guess at, compete for, and co-create jointly built narratives of loss. I then examine how these narratives are aired for a televisual public via displays of grief that produce a kind of 'takeaway' trauma: infinitely consumable, ready-made, disposable and cathartically satisfying. I argue that while Edward's ghosts operate as social indicators of loss, they are rather produced by a process that enacts the loss of loss. When this constitutive frame of lost loss accidentally appears, it creates a process in which *feeling is used to remediate itself*. As a momentary undoing of the performativities of the memory affect, this exposes the experiential dimensions of *memory's affectivity* and reveals the crisis of the ghost world that Edward works hard to contain.

Thus, it is in the transit between the live and the televisual that *Crossing Over* engages and produces the memory affect: the televisual remediation of the live and the live's remediation of the televisual. It produces the quality of *feeling, remediated*, which explains how memory affects, as formations of liveness, are sensuously contaminated by the re- and premediating principles of a televisualised sociality. This project of feeling memories is particularly invested in positioning the primacy of the televisual as an originary sensory source, which, at the same time, produces its own televisual afterlife. The twin acts of remediating feeling and feeling remediation incur memory as a trace, or echo, of itself. As I argue at the close of this chapter, these dynamics of televisualised memory can be most clearly seen in the controversial banning of Edward's themed '9/11' episodes, which were produced in the weeks following the September 11, 2001, attacks on the United States, and were swiftly prohibited from being aired by the TV network. While Edward aimed to draw on the sentimental capital of September 11 as a dominant trauma referent, the banning of his episodes reveals his failure to understand the event of September 11 as already televisualised, that is, as a response to, and manifestation of, the very form of sociality generated by his own televisual apparatus. Through what the banning of these episodes unwittingly makes appear, I uncover the memory affect as a kind of spectral backdrop to postmodernity itself: the ghost of a ghost. In mobilising the ghosts of ghosts, it engages a *felt absence of feeling*, which is incurred in moments that collapse the normative dimensions involved in everyday ways of feeling the memories of others.

Canine clairvoyance

> Edward (continuing on the subject of Dano's grandmother): 'I'm getting "Bo-bo." '
> Dano (perplexed): 'Bo-bo?'
> Edward: 'Like two b's. "b-b" ' (Dano is not making connection with any human she knows.) 'Wouldn't be a dog, would it?'
> Dano: 'Beebee?'
> Edward: 'Passed?'
> Dano: 'Yes!' (Dano pauses and a look of amazement crosses her face.) 'No! I had a dog named "Beebee"!'
> Edward: 'Passed?'
> Dano: 'You get dogs?'[6]

John Edward *gets* dogs. In fact, Edward says that he can communicate not only with dead animals, but also with dead famous people, dead historical people and even with dead babies. 'What if we could have a snapshot of all the people John Edward said were clamoring to send a message to the living?' asks sceptic John Hockenberry. 'Well, they're not angry folks. They're not bitter at being taken too soon. We heard no stories of pain and suffering, no calls for revenge or settling scores. These dead people are *nice*. And maybe a little boring.'[7] Even though Edward's subjects have died gruesome deaths, fought family feuds or suffered catastrophic accidents, once in heaven there is peace and tranquillity for everyone – including the family dog. The image of Edward's pet heaven is not only bolstered by his syndicated television programme, but also by his live seminars, private readings, appearances on *Larry King Live* and numerous DVDs and self-help books. His most recent book explains how he became what one reviewer describes as 'one of the few growth industries in an otherwise lack-luster economy'.[8] Although now a global franchise, Edward's story began as a 15-year-old boy in moments where he 'knew things' that he 'shouldn't have known, family events that...no one had told me about'. He refers to his spiritual guides – 'the boys' – as keepers of his sixth sense: 'They were leading me to the understanding that I was on a path to a life's work connecting the physical world to the spirit world.'[9] From the development of psychic skills, to debuting at psychic conventions, to appearing on talkback radio, Edward sees the irony in his media mediumship: 'I can see dead people' is a catch phrase that not only he can exploit.

Edward is uncannily aware of his double identities as a media icon and psychic medium, characterising his work as a 'blend of spiritualism and entrepreneurship'.[10] His commercial success is demonstrated in *Crossing Over*'s overnight ratings high, with *People Weekly* reporting that since premiering on the Sci-Fi Channel in 2000, it became the most popular programme in syndication, with an audience of more than 3 million – making Edward into a 'whole other sort of astral projection: a TV star'.[11] The tabloid press are tentative about his handling of delicate subject matter, with critics variously debunking *Crossing Over* as using eavesdropping, planted actors and cold-reading strategies in ploys in which a medium delivers a stock questionnaire to persuade an individual that he knows all about their problems. *TIME Magazine* ostensibly uncovered Edward's scamming, accusing him with charges that 'aides were scurrying about, striking up conversations and getting people to fill in cards with their name', which were 'picked up by the microphones strategically placed around the auditorium'.[12] Edward's standard response to such allegations is to replace the terms of provability with morality: 'what does get me upset is when people try to define my motivations... [s]aying I'm in it for the money or that I'm taking advantage... or that I'm a phony'.[13] While Edward justifies his spectral economy with terms that bolster its integrity, the business of *Crossing Over* might also be found in the question of what his ghosts have been exactly conjured to do. Here, we understand, as Avery Gordon points out, that the ghost 'is not simply a dead or a missing person, but a social figure... that dense site where history and subjectivity make social life'.[14]

Historically, ghosts are cultural disturbers. As the spectre of an often gravely unhappy end, ghosts symbolise morality and revenge, doomful fears about God and the hereafter and fits of conscience. As cultural actors, ghosts perform a 'becoming-visible' and embody that process in ontological form. In this sense, they are predicated on a conflict in the term's double entendre, where 'apparition' inscribes an act of becoming (appearing) and a state of being (or seeming to be). They are, in this way, as Gordon explains, constructions *of*, and *with*, affect:

> Being haunted draws us affectively, sometimes against our will and always a bit magically, into the structure of feeling of a reality we come to experience, not as cold knowledge, but as a transformative recognition.[15]

It is both how ghosts mechanise appearance and *appear to be* that reveals their relevance to a consideration of the ways that memory cultures produce practices of cultural memory. Further, and as I will argue here, it is the particular 'structure of feeling' engaged by Edward's ghosts which masks the programme's other hidden effects. That is, as processes of feeling pasts, Edward's ghosts are cultural formations of the present.

While ghosts are cultural disturbers, postmodern ghosts are no ordinary spectral assemblages of this ilk. As John Potts suggests, via his notion of the 'ghost idea', the contemporary moment witnesses the probable erasure of the ghost altogether: the force of liveness upon which the ghost has relied has been made redundant by the digital field. Ghosts no longer exist because their paradigms of disturbance are no longer facets of the mediatised sphere: haunting only happens by way of informatic data flows or cinematic effect.[16]

Gordon observes that the characteristics informing a so-called ghost-less society include

> television-structured reality, the commodification of everyday life, the absence of meaning and the omnipresence of endless information, the relentless fascination with catastrophes, and the circulating advertisements for the death of the author, referent, and objective reality set within image upon image of the electric connections among life, death, and sex.[17]

At the same time, however, Gordon warns against the false positivity of arguments that simply erase the authentic mystic paradigm's traditional capacity to haunt. Postmodernity, she suggests, does not so much erase ghosts, as perform a condition of erasure to mask what is a candid presence. This might be understood as a kind of 'double-ghosting' set forth by what sits underneath the loss of modernity's access to the real: as the cultural production of the loss of loss, which was identified in this book's introduction. In this sense, the death of the ghost as given by the ghost idea might rather position Edward's ghosts as uncanny returns of the technologically repressed. If ghostliness can die, then Edward's ghosts can be seen to materialise specific forms of haunted return.

As Potts explains, the technology responsible for the evaporation of the ghost was originally tied to its historical evolution as an entity that stages ontological uncertainty. The contemporary displaced ghost hence arrives with firm past precedent. Correlations between the supernatural and the technological can be seen in the drives for disembodied communication witnessed by industrial modernity, and in particular, in the way ghosts appeared alongside specifically new innovations for communication. In early spirit photography, for instance, crossovers between medium and media were made visible by practices that tricked the light-sensitivity of the form. Transparent bodies would float above the stiff poses of mid-19th century life: vague ghosts, sad-looking ghosts, ghosts as skeletons and ghosts as ectoplasmic glue. Family portraits and the strange, stiff postures that arrive at an historical turning-point reveal the correlation between the drive of technology to capture life and the drive of mysticism to undo oppositions between body and spirit, life and death. These drives were also witnessed in the late 1800s' paranormal complement to the arrival of the telegraph – the rise of Spiritualism, a cult communication with the dead. Identified as the first technology enabling the flow of disembodied data, the telegraph broadcast messages across wire, a revolution in the bodily ownership of the word. Spiritualism similarly channelled the deceased through imitative rituals of 'tapping', which used séance to counter – but also meet – the new sciences of modernity. As Jeffrey Sconce notes:

> While the technology of the telegraph transformed America into a wired nation, the *concept* of telegraphy enabled endless displacements of agency, projecting utopian possibilities onto a disembodied, invisible community and recasting an often radical political agenda as an act of supernatural possession.[18]

As these tandem histories suggest, it is in the crossovers between digital and supernatural, and their implications for the construction of a broader public imaginary, that an emergent politics surrounding the matter of ghosts can be perceived. The ghosts of modernity were not merely remnants of the past, but effigies ghosting variable crises in the practices of self and polis amidst dense social and industrial change. That the conceit of mobilised agency signalled by early telegraphy was symptomatised by Spiritualism prompts a reading of the contemporary televisual apparatus as not only beset with spectral qualities, but also with a similar agenda evidenced in the ghosts that it constructs. While part of Edward's psychic platform is the extraordinary ability to 'connect' with non-verbal entities such as a dog, the cultural imaginary that he builds confirms the superiority of his psychic capacity and stabilises the picture of a secure pet heaven (within a secure human afterlife) that carries the forms of normative subjectivity Edward works hard to create. Edward's ghosts use a postmodern mechanics of appearing to stage a pre-modern vista of appearance, and both are central, in combination, to the cultural politics of recollection that the programme assembles. Here, Edward's imaginary – a stabilised pet-friendly ghost heaven – occurs at a pivotal cultural moment: the very time that sees the erasure of the ghost altogether. As distinctly postmodern ghosts – televised reappearances of the televisually repressed – they are a curious cultural paradox materialised in none other than Beebee the ghost dog.

Spectral semiotics: The letter disguised as a ghost

Crossing Over is a pre-recorded televisual transmission of a live performance that occurs between a medium and his client – a practice that Edward claims he's 'been doing for years... with just God's help'.[19] The dramaturgical mechanics of Edward's live mediumship – from his charismatically charged, embodied psychic performance to that of his participatory audience members – involves using the memories of some to generate the memories of others. Below I describe in detail how the felt, sensorial capacities by which Edward's tracing-giving process occurs creates a televisual sphere of sentimental effects and affects which arise as cathartic 'punchlines' for home-viewers. As one facet of the broad terrain of memory cultures traversed in this book, Edward's memory affects foreground a correlation between the embodied reproduction of traumatic pasts in the 'televisual present' and their ongoing, commercial afterlife. In short, Edward's live studio audience are sensuously complicit in the very mechanism that works to produce their own televisual commodification.

As a live event, Edward's form of mediumship involves the narration of memory traces that link spirits of deceased relatives to audience members via a process of data validation, given in the calling out of information that he 'gets' from 'above'. He firstly receives a message in the form of a letter as a trace – a sound that registers as, for example, 'two b's', a 'b-b' or a 'bo-bo'. Edward then locates the area within his audience that the clue might be aimed towards. Clues usually begin with letters that signify names, but can also be in the form of numbers interpreted as special dates: a birthday or anniversary, for example. Other clues can involve the description of a death, an image of a personal item or a spoken phrase. A spectator's role within Edward's production of data traces is to 'validate' those traces. Validation ideally occurs with simple negative or affirmative responses, but usually occurs with segments of elaborated information that bolster the otherwise uncontextualised trace Edward has provided. From the outset, Edward's narration of the trace engages a performance of recollection, which is particularly charismatic. He pauses with a quizzical expression on his face when 'receiving' data. If this trace is numerical, he will factually communicate it. If it designates some form of experience (most usually the experience of the death being communicated) he will express it through acting out its imprint on his body. Here, he may gesture to specific parts of his body, or explain the mode of death through a felt correlation with that of the person 'who has passed', as in the excerpt transcribed below: 'there's an inflated feeling coming within side me so I don't know if a lobe of her brain started to swell'.[20]

In the first stage of his live mediumship, Edward establishes the existence of the ghost world through his direct conversation with and embodiment of it. In the latter, what spectators witness is Edward re-feeling the traumatic meme itself: the moment of death is affectively reproduced through its difference from, but reliance upon, Edward's body as a 'channelling' conduit. Here, Edward establishes his star status as a kind of ultimate remembering body, one who can sacrificially re-feel the traumas of the past in the name of healing those in the present. In one sense he literalises the kind of prosthetic forms of recollection that Alison Landsberg theorises, particularly in how his performed gestures choreograph the forms of the co-embodied 'suturing' Landsberg discusses. To repeat her theory here: 'prosthetic memories... are not the product of lived experience, but are derived from engagement with a mediated representation... and like an artificial limb, they are actually worn on the body'.[21] While Landsberg's study makes clear that prosthetic memories evolve as an experience of mass mediatisation, Edward's body instead acts as the media interface itself. In this he becomes a public archivist of histories of traumatic death, promoting the capacity for those memories to function as culturally re-embodiable memes that anyone might re-experience. Through this, the twin modes of narrative and feeling that he crafts become essential to how he produces memory traces that will then work into the bodies of his studio audience, which in turn become ghostly 'structures of feeling' for home-viewing spectators.

As the next step in Edward's live mediumship, the trace-summoning process involves the whole audience in the validation of his ghost world. The pause in the excerpt above, between Dano's confusion over 'b', and her later acknowledgement of 'b' as 'Beebee', indicates how the central ambivalence of the ghost traces work as a suspended referential leap during which the entirety of the audience can claim the memory signifier Edward emits. This moment is essential to understanding how the trace generates the memory affect – the *production of the feeling of a trace of feeling*, occurring as a new memory formation in its own right. Through it, we embody (or are co-produced by) the feeling-*ness* of a particular past, one that is remounted as *that which ought to be re-felt*. Here, what Edward has already 'felt' of the trace dislodges from his own body to become a potentially re-feel-able meme that his audience work hard to corporeally place. For the duration of the time that the trace remains unclaimed, the audience filter their own memories for a match. It is here that projecting onto a trace provokes the possibility for the trace to become a story. As a collective force, the audience support the deictic power of the trace through shared desire. As narrative builds into the trace, Edward feeds in further traces to locate a singular audience member in connection to the ghost and to expand the moment of embodied mediumship into a moment of confessional story. This is a moment where the sensorial attributes of the ghost as a feel-able past transform to produce its sentimental afterlife: the memory affect materialises to political effect.

As I explained earlier, a postmodern ghost is one which *appears* as a ghost to mask that which haunts the broader commodification of everyday life – the absent meanings and relentless fascination with catastrophes that Avery Gordon sees as being constitutive of contemporary 'television-structured' realities.[22] This masking effect becomes evident in the oscillation the programme strikes between establishing a stable ghost world on the one hand, and the highly unstable process by which it establishes this ghost world on the other. It is exemplified in how Beebee the ghost dog exists as both a thing that is named and a linguistic repetition that is sonic but not meaningful in itself. In being a sound pattern and a lost object, Beebee signifies signification: she is a random sound pattern produced to make meaning, and simultaneously, the resultant memory-effect of a referential sequence. As a memory trace, Beebee collapses the process of enabling appearance (a linguistic game) and the desired result of appearance (a ghost) within her singular but complex ontology. In being a name and a phonetic letter, Beebee alphabetises herself into being. Thus 'b' becomes 'Beebee' becomes a lost family dog. As the audience wait for Edward to confirm that 'b' is really Beebee, and as they then wait for Dano to confirm that yes, there was 'a dog named Beebee', *Crossing Over* articulates its ghost practice as the coincidence between an arbitrary practice of naming and a traumatic practice of mourning.

Performatively speaking, Beebee is both the process of naming and the after-effect that the naming produces: a phoneme cleverly disguised as a ghost. What Edward is playing at, as much as memories, is the semiotic system: meaning-making itself.

The certainty that arrives with the playing out of memes of signification can be further explained by theorisations of affect that connect potentialities of feeling to the way that signs perform cultural power. While beginning as arbitrary letters or traces, the signs in Edward's work operate to become readable signs of trauma; they perform an affective potential when not being narratively interpretable, and they produce other affect-effects when they do finally materialise narratively, as ghosts.

Brian Massumi and Sara Ahmed observe how signs and objects create feeling bodies as well as how they circulate feelings about bodies. Ahmed relies on Nietzsche to consider the 'pain affect', in which an object comes to refer retrospectively to the feelings that it can cause. She cites Nietzsche's analogy of the symbology associated with a nail, such that 'I can just apprehend the nail and I will experience a pain affect, given that the association between the object and the affect is already given. The object becomes a feeling-cause.'[23] In Massumi's Piercean reading, signs work their felt effect on the body in a similarly anticipatory way. With the analogy of a fire alarm that intrudes on a sleeping body, the body is set into motion by the affective *perception* of threat (a fire). As Massumi explains: 'The sign is the vehicle for making presently felt the *potential force* of the objectively absent.' It is 'a question of how a sign as such *dynamically* determines a body to become, in actual experience... how an *abstract force* can be *materially* determining'.[24]

In view of both analogies, the reanimation of signs of trauma in Edward's mediumship contributes to a process of affect-creation, whereby a series of memory traces determines the sensibilities, and communities, which grow to surround circumstances of narrated trauma. As signs that perform a becoming-ghostness, Edward's psychic traces play out this transition from abstract force to material determination. Further, where the successful appearance of a ghost is brought about by the fortuitous correlation between a participant, their narrative history and the stories that Edward narrates, Edward relies upon this correlation to produce an ultimate home-viewer effect.

During the reading process, the audience in Edward's live gallery, like Dano, negotiate positions of viewer and participant. They witness readings conducted for other audience members and anticipate readings that may be conducted for themselves. While these dual positions require different techniques of performance and reception, they also place viewers in differing subject positions in relation to the economy of the programme. A spectator can expect to narrate, remember, guess, watch or confess, for example. This is not simply a signposting of varying responses to the programme, but is much more significantly a process of playing out relationships to the cultural commerce of recollection itself. In this way it could be said that Edward appropriates generic signs of trauma as a way to mobilise particular 'feeling-causes' in his live studio audience, who then become ghost objects or 'feeling-causes' for home viewers. In doing so, he draws the 'pain affect' carried by tropes of trauma onto his own body and reanimates it outwards, producing a kind of pre-purposed state that is to be named by spectators who materialise the memory that it produces. Here, spectators reanimate a historical trace for a globally televisual public. In laying claim to a trauma affect, they draw the *sign* of trauma into repertoire, embodying its resonance as a site of affective power. This self-conscious engineering of affect is what is harnessed for spectators at home.

If we understand via Massumi that the sign has material implications even when it works as a threat (as in a *false* alarm), then Edward could be said to engage spectator bodies by drawing on the productivity of the threat: producing the affective perception of what might never eventuate as a means to cultural power. The memory affect is animated in the way that home-viewing spectators view enactments of memory that are simultaneously pre-empted by, and exist as after-effects of, televisual time and space. Via his traces, Edward produces 'feeling-causes' which pre-state what they will materially determine, 'making presently felt the *potential force* of the objectively absent'.[25] Via Richard Grusin, we might say that Edward's ghosts are premediations of television's central affectivity.

In Grusin's theory of premediation, televisual logic is premonitory in the sense that it is fed by threats that it would also perform itself as warding off. While for Grusin, remediation is the reframing of one medium by another medium (a concept I turn to in

Chapter 4), premediation instead envelops the mediations of the late 20th century within a temporal machinery characterised by terror. Here, Grusin notes that the 'shock' of the September 11 attacks on the United States generated a radical departure from mid-1990s televisual immediacy to incite a culture of precognition. Premediation is located in 'real-time technologies [such] as video and the Internet' which favour 'a prophetic or predictive role of reporting on what might happen'.[26] In this, the televisual envisages events before they happen: news media don't merely get there 'first', but get there 'before'. In being anticipatory, they create a low-level social panic and also produce the salve for that panic by warding off the very threats they have created.

While Grusin uses premediation to speak to the themes of security particular to reality television, the logic of premediation can also be seen in the mechanism of liveness that *Crossing Over* generates. In effect, it uses the past to anticipate a formation of the live that it would also aim to contain. While remediation argues that the televisual repurposes the live, premediation is predictive such that it produces a formation of the live that has already formed itself as a conception of the televisual. The prophetic role of the televisual emerges as that which understands *the role of the live in anticipating its own preformation as televisual*. These oscillations can be seen in how Edward's live performances anticipate the televisual, and in how the televisual uses a form of liveness that is pre-construed for these purposes. Edward's studio audience thus 'premediates' itself as a function of the televisual, which means that when their stories are transmitted for televisual display they arrive uncannily ontologically pre-prepared.

As an abstract force that becomes materially determining, Beebee the ghost dog characterises the way that the televisual logic that unfolds in Edward's programme consists of a liveness that is not only premediated, but is also premedia*ting*, in the sense that it operates to ward off future threats. This anticipation of the future is, as Grusin would argue, self-justifying: the ostensibly predictive aims of premediation are also, insidiously, formative. This is apparent in the links he makes between premediation as a preventative mechanism, 'in which the United States seeks to try to make sure that it never again experiences live a catastrophic event...that has not already been premediated', and as a reactionary politics. In the second, we see that the pre-emptive media logic of the Bush administration's doctrine of warfare was also chiefly enabling of that warfare:

> In a political regime of preemptive war, premediation became the dominant media regime – by premediating the war, remediating it before it happens, the formal structure of U.S. news media effectively supported U.S. military doctrine, participating in the preemptive remediation of a future (premediated) war.[27]

If the post-September 11 media generate those very realities that they would otherwise prevent, then the logic of premediation operating in Edward's programme can be perceived in the way that his ghosts construe the very open-ended, unresolved histories that they are also summoned to resolve. While a media logic that enforces pre-emptive warfare imagines terror as a self-justifying threat, Edward's televisual logic works functionally rather than thematically, premediating those threatening forms of sociality that it would otherwise aim to contain, and promising instead to generate forms of subject-certainty via what it pretends to ward off. In other words, *Crossing Over* seeks to ensure that the public never witness a form of spectator subjectivity that has not already been premediated, which is to say that it ensures that its participants only ever construe themselves *through* the affect-mechanisms of memory and self that the programme produces.

In this sense, the boringness that is central to Edward's supernatural, typified in the image of a dead family dog, can be understood to signify a world that ghosts a much larger crisis – not a crisis of death, but a crisis of obtaining subjectivity outside of the re- and premediated affective circuits that produce it. This larger crisis destabilises those very practices that aim to govern subjectivity as concretely unified and whole. In this, Beebee is a motif that reveals why *Crossing Over* needs to produce a highly formulaic ghost world in order to hide the play of chance that it enacts across the socio-semiotic field. The mechanics by which Edward is able to make his live programme work, and also by which he is able to televise his re-edited programme to a home-viewing audience, are made material by Beebee's doubled ontology as an 'apparition' who is both appearing, and appearing to be. The connection, however, between Beebee's cultural identity (her ghost effect) and her ontology (the game of appearance) is here pointedly marked. We begin to see the game like logic that connects a felt semiotics of loss to a culturally emotive thematic of loss. While the mechanics of appearance are built around endless combinations of affectively charged signs, these nonetheless stage real memory effects. While Edward's audience want contact with what they have lost, what they do is enact the loss of loss itself. Between the two, they remediate each others' feelings as well as *feel* this process of remediation, which, as the concept of 'feeling remediated', takes the shape of the memory affect.

The cultural afterlife of sensing the afterlife

Edward's gallery is intensely serious. The episode channels a teenage girl who died suddenly from encephalitis – viral swelling of the brain. Edward 'connects' her with her grieving parents:

> John: I'm supposed to acknowledge for you that there's surgery that happens in the head area, okay?
> Parents (crying): Yes.
> John: I want you to know that I know that you're thinking that something went wrong, okay?
> Mum (crying): Yes.
> John: I know that you're thinking that.
> Mum (crying): Yes.
> John: But you have to understand that the type of surgical procedure that was being undertaken was huge, okay? It feels huge to me and very, very delicate. And I feel like what happened with her was explained to you in some way. I don't know if this was like an emergency surgery, I don't know, but I feel like I need to stress to you, I feel like it's explained in detail: these are the precautions, these are the things, the negatives... It feels like it's a freak thing that takes places within side the surgery itself; a reaction to the procedure, a reaction to the anaesthesia, there's an inflated feeling coming within side me so I don't know if a lobe of her brain started to swell, okay, but there's an inflation that they're trying to show me (parents nodding, crying). The reason why I'm being this graphic is because she's being this graphic, and the reason why she's being this graphic is to validate for you both – (to the father) I don't know if you're Dad – but to validate for you both that she is okay.[28]

I watch this episode on television. In the televised post-reading interview, the parents describe the psychic experience as the bittersweet quality of 'going to visit somebody that you haven't seen in a long time and knowing that you're never going to be able to see that person again'. The studio audience appear to be still, even stiff. It seems that even if they are not convinced that they are in the presence of an actual ghost, they know that they are in the presence of trauma; this, and the kind of story every person would dread to have made their own.

In this example, the cultural politics of feeling memories of others surfaces in the relationship that occurs between Edward's live studio audience and his home viewers. It is in fact the televisual frame which constructs the discursive space by which Edward can perform traumatic cultural memory in the embodied way that he does, and further, it is the televisual's demand for the affectivity that emerges from his recorded liveness that drives the mechanics of his studio format. As one factor within Grusin's thesis of premediation, reality television has been noted by Jon Dovey for its successful 'institutionalised blurring ... between performance, mediation, narrative and fact', which has now come to signify a new age in televisual practice.[29] Within it, the trope of authenticity has spurred a diverse range of surveillance, documentary, chat show, competition and docudrama genres that now propagate the excessive visibility characteristic of the broader mediatised field. This is the kind of visibility that, as Nigel Thrift argues, is enacted as the 'the performative principle at the heart of modern Euro-American societies and their political forms', which, present in the interminable screen, 'foreground[s] emotion, both in its concentration on key affective sites like the face or voice and its magnification of the small details of the body'.[30] *Crossing Over* markets the affective potential of the screen by focussing on visual markers of emotion and trace-markers of traumatic histories, and, as such, it could be seen to be the ultimate manifestation of late capitalist memory culture's affect-driven media apparatus.

As Dovey argues, within the multiple genres of reality television, each shares a predilection for making intimate revelation a 'key part of the public performance of identity', and importantly, each mode is decoded by viewers with an acute sensitivity 'to precisely [the] kind of reality reference any given show is based upon'.[31] *Crossing Over* in this way relies upon the pre-training of its televisual audience for its popularity. This involves the production of its reality reference (the ghost) through the internal genre referentiality the programme establishes. Here, while it is primarily a psychic TV show, it is also identifiable as a chat show through the way that Edward responds to an interactive and expositive audience. Further, it is identifiable as a form of trauma TV in which 'individual tragedies which would have once remained private... are now restaged for public consumption'.[32] Finally, it is identifiable as a live competition in which studio audience members compete to have their stories of grief psychically read. Each format works to stabilise the programme's reliance upon another format, and put together, the mélange of formats makes for an exceedingly consumable product. This genre referentiality is taken further when *Crossing Over* references external televisual events. In one example, Edward devised a themed special to revisit his 'favourite' daytime soap opera, *The Guiding Light* – itself based on a ghost plot. Viewers could 'tune in all week as [their] favourite celebrities [actors from *The Guiding Light*] connect[ed] with their loved ones and share[d] incredible stories... that out-soap the soaps'.[33] Here, Edward gave psychic readings to the cast of *The Guiding Light* to produce *Crossing Over* as a 'real-life' psychic soap opera.

Crossing Over's mixed and crossed reality references are built upon an interesting inversion, but utilisation, of the reality form. As such it uses the conventions of reality television to play at multiple truth regimes simultaneously. While it appears to be a programme that uses a reality TV format to channel paranormal content, it is rather a programme that uses the pretence of the paranormal (the seething presence of a 'seething presence', if you will) to force a reality television result. This kind of result can be thought of in terms of what Laura Grindstaff has theorised as the reality TV money shot,[34] which draws the spectatorial economy of Edward's live audience into the immediate accountability of his home-viewing audience. Drawing on theories of pornographic performance and spectatorship, Grindstaff argues that the money shot, more usually conceived of as the moment of sexual climax, can be characterised in reality TV as the moment when personal trauma becomes televisual spectacle – when seemingly ordinary people emotionally explode. These are understood as instances when studio guests lose control and express joy, sorrow, rage or remorse on camera. Fans contend that what makes Edward uncanny is 'how he nails down concrete details from wispy hints',[35] and how he can 'crumple audience members into hysterical blubbering by giving proof – through dates, names or remembered objects – that their dead relatives were saying hello'.[36] In view of Thrift's argument for how the televisual works to produce affect through the highly charged site of the face, Grindstaff's notion of the money shot accounts for the role of trauma in the production of these effects as their own kinds of after-affects, as Jim Scurty's (Edward's cameraman) narrative makes clear:

> I was shooting the close-up and watching the expression on his face, watching him begin to fall apart. First the tick and then the trembling in the chin and the kind of embarrassment, the eyes shifting while seeing if anybody's watching. And I literally watched this person fall into this grief and surrender to it publicly. It was very hard to watch.[37]

Edward works on the premise that he conjures ghosts in order to heal – in his words 'to validate for you both that she is okay'. From a performative viewpoint, however, his ghosts appear for their money shot capacity: they produce consumable trauma. In this, the pre-recorded live moment of confessional grief is further bolstered by additional memorabilia. This occurred in the episode transcribed above, when the programme visited the deceased child's parents to film her neatly preserved bedroom. Face-to-face with the after-effects of death, the home viewer receives a picture of a life compiled through a now public personal archive. Televisually, the deceased person becomes a reconstructed object, an ironically produced after-event. Live studio participants who 'begin to fall apart' become premediated affects that perform a televisual afterlife.

Dovey's analysis draws correlations between the reality genre and late capitalist investments in the normativising capacities of image and affect-making regimes. He argues that the traditional documentary form has suffered: having once been invested in a democratic public sphere of 'truth' it now uses an electronic super-space based on an unprecedented inversion of public and private. This is partially owing to the ways that neo-liberal economies remove the resources that would support more balanced or nuanced documentary texts, cheapening production values and accelerating outputs over content in the process. These kinds of formats further signal an inversion of the role of the televisual, which, as a cultural memory mechanism splinters subjectivities and promotes unified subjecthood at the same time. He writes:

> [t]he significance of Reality TV hybridisation lies not in the way it signifies either economic imperatives or postmodern genre collage but in the way it insists on the primacy of the individual, emotional, and above all unified version of subjectivity.[38]

What was once thought to structure a collective communicative space now structures the very ethos of that space, but in simulation. Reality TV, as pseudo-documentary, maintains a pretence of democracy but reinscribes more hidden forms of disciplinary control. For Dovey, this confirms what all reality television genres aim to generate: 'the production of normative identities' which restore 'lack of narrative coherence' so that '[s]ubjectivity, the personal, the intimate, becomes the only remaining response to a chaotic, senseless, out of control world in which the kind of objectivity demanded by grand narratives is no longer possible'.[39] In this sense, while the televisual version of *Crossing Over* exposes an excessive postmodern hybridisation of form, this kind of hyperbole smoothes over the multimodal nature of the programme to instead promote images of a unified ghost world. To this degree, audience members unknowingly maintain the programme's disciplinary system, which operates in the service of its (hidden) system of chance.

The twin ideas of a particular reality reference (the production of an exceedingly simulated appearance) reproducing a particular regime of truth (the production of secure and stable knowing) emerge in *Crossing Over* in the way that participants perform themselves as distinctly fragmented against, and in pursuit of, that which is perceived to be whole. Collage is used to produce its opposite: a performance of unified subjectivity seen in grieving parents, an empty bedroom, a young girl's tragic death. Such images reveal how *Crossing Over* continually uses ghost affects to reconstitute the spectator as an agent of certainty in relation to systems of authority that specifically exceed their control, but that are designed to give the *impression* of each individual having control. While spectators appear to be attaining a connection with ghosts, *Crossing Over* works to display them in their continued performance of the attainment of subjectivity itself. It is this performance which produces the memory affect, and which further reveals its cultural effects. Enforcing the regime of truth, then, is a regime of power by which participants attempt to attain discursive control over a system that is in effect always already defunct. Across this process, we witness an affective 'spill' as subjectivity unravels and is remade, as participants work to generate a collective ghost pool, and refigure themselves as the primary ghosts of Edward's vision. This process of 're-deconstruction' importantly occurs on and through their bodies working as collective memory agents. In their failure to register as singular and complete, Edward's audience instead pursue – in the image of a secure ghost world – their own discursive deaths. Rather than its ghosts, this is possibly the biggest haunt *Crossing Over* constructs as a televisual social sanctum.

Feeling losing loss: The mechanics of a 'me-too'

As the Beebee example illustrated, Edward enables a ghost to emerge in the relationship between a memory trace held by a viewer and the mnemonic effect generated by the calling of linguistic structures, such that a 'b' becomes 'Beebee' becomes a lost family dog. While the main task of *Crossing Over* is to secure this referential leap in order to produce a money shot, Edward's role as convenor of this process is to manage the flow of memory traces between mnemonic triggers and audience identities. This flow itself is what uses, and mobilises, affect. It is also a process that interestingly fails when audience members cannot align with the traces Edward provides. In situations such as this, the collective process of generating a ghost breaks down and the apparition becomes invalid through the conflicting validations it acquires. These are importantly instances in which the ghost, as a secured and stabilised idea of appearance, begins to disappear. It is no longer a ghost; it is not even that. As an uncanny showman, Edward's phrase for justifying these slippages is a 'me-too': moments in which multiple ghosts appear at once, confusing Edward as to who is what ghost, belonging to what family, speaking to what audience member. This can also be read as a moment in which varied traces and their respective audience recipients are in competition for the cultural power or force associated with an 'affect cause', leaving the errant memory signifier suspended without chance of referential closure until it is singularly claimed:

> John: The first person I'm coming through to is a male figure to the side, a husband or brother. (To Andrea) You're saying your brother is passed?
> Andrea: Yes.
> John: I'm going to tell you that your brother, if it's him, is bringing through an older male, so your father, your father-in-law or there's another male figure above...
> (Andrea cannot acknowledge this).
> ...
> John: (Voiceover) At this point it seemed these messages were for Andrea and her husband Dan. However, sitting right in front of them was another family and judging by their reactions, I could tell that Jean and her daughters wanted to join in. This reading was about to become a classic example of a 'me-too', where two or more different families have loved ones coming through together.
> John: (to Jean) Are you connected to the Joe?
> Jean: John, my father's name is John.
> John: Woah, hold on a second... I have the male figure to the side who has to be through first, who wants me to acknowledge that the father figure is with him there. There is a J-o name like John, like Joe, I thought it was Joe. And there is a lung cancer, emphysema, blackness to the chest, connection and... the third month, March, or the third of the month is connected to this in some way. (To Andrea) Did your brother pass from a suicide?
> Andrea: My brother? No.
> John: There's also someone who passed from a suicide where their actions brought about their passing. What they did caused how they passed.
> Andrea: My friend was in a car accident.
> John: No, no, no, this isn't a car accident. This is either somebody who either accidentally electrocuted themselves... there's an accidental type of... but they take on the responsibility on for how they passed.
> Andrea: I had a cousin.
> ...

John: I'm not really sure who's coming through, or which family I'm talking to, to be quite honest. I don't know if it's your family or your family...[40]

In this excerpt, Edward first introduces Andrea's 'brother figure' who is also in the company of (Edward uses the term 'bringing through') a 'father figure'. Andrea cannot validate the clue of the father figure and so her ghost then becomes Jean's ghost and is, in this confusion, potentially not Jean's ghost either – it is strangely both or neither, and the entity which registers 'a J-o name' and a 'blackness to the chest' hovers above the room as an unknown spectre waiting to be claimed. And yet it is the desire of the audience members who attempt to match the data that keeps the possibility of appearance intact. Andrea's response to Edward's description of a suicidal death with the declaration that her 'friend was in a car accident' reveals the struggle for identification audience members push to experience. As television critic Josh Wolk observes of another reading:

Edward was getting a K name who died from a bad blood transfusion. When no one spoke up... I felt those around me struggling to remember a connection that could help them claim it. 'It's kind of a psychodrama where there's a willing suspension of disbelief... All the players are integral to it'.[41]

The significance of a me-too not only rests in how it reveals the building of an accepted ghost world, but also how it reveals what spectators rehearse of their own affect reservoirs or memory banks as they become tied to the ideologies of appearance created by this process. While the aim is to reach a stable image – a fantasy ghost world – the process of arriving at that image conflicts with the establishment of that world: a process based on performing reference in crisis. For Andrea, it is neither completely her brother nor her father, nor not her brother nor father who communicates. Instead, the most direct communication exists between herself and her contender, Jean, as they try to affirm themselves through the ghost traces they are given. At the centre of this me-too is the fraught proposition that, for a participant, the stability of a subject position promised by the programme is continually undermined by a central ambivalence surrounding ghost ownership. This can also be understood as a competition for dominance of the affective potential of the social semiotic space. What Andrea and Jean want to lay claim to is the felt traumas of a single letter 'J' – a letter disguised as a ghost – and its attendant promise of discursive power. Instead, Jean and Andrea become witnesses to their own negated feeling in the very moment that the phoneme 'J' will not materialise as a ghostly effect, but rather materialises their desire in play. When faced with losing the letter 'J' and its promise for generating a narrative of loss, what Andrea and Jean feel is rather the loss of loss itself. As the appearance of the programme's truest ghost, here emerges the ghost of ghostedness: a *felt absence of feeling*, incurred as a collapse, or implosion of the normative dimensions involved in everyday ways of feeling the memories of others.

As I discussed in the introduction to this book, the memory affect appears as the spectre of memory cultures which aim to restage loss but inevitably re-lose it. As I have shown above, Edward's game is deeply invested in the mechanics of producing and retaining loss, a premediating mechanics that is nonetheless destined to fail. Amidst the commodifying machine that is Edward's programme, the moment of the 'me-too' reveals that there is, as Gordon suggests, a vitally important 'structure of feeling' to be beheld in moments when the machinic components of psychic TV breakdown. Here, a re-perception of the loss of loss might be provisionally found. I canvassed this idea in Chapter 2, when the staged scenario of the colonial encounter at Tjapukai Park momentarily fell apart. In that discussion, the collapse of a fire-lighting spectacle managed to ignite a re-perception of the socio-historical dynamics carried into the present by that very stage apparatus. In a very large sense, the 'ghosts' in that space were the traces of all bodies complicit in that colonial history, across time. Here, Edward's ghosts use late capitalist culture's investment in the pains of others to disappear its polarising investment in everyday bodies and selves. The hauntedness sitting underneath this successful ghost machine surfaces, ironically, in the moment that the fiction of the ghost collapses, and in this collapse there emerges another form of affective instability: the *ghost* sitting underneath his ghosts. In *Crossing Over*, the appearance of memory's affectivity at precisely the moment when the ghost idea unravels reveals the often-overlooked truth-effects mobilised by playing at feeling the memories of others.

Ghosted ghosts: The missing episodes

If losing loss causes an affective spill to enter the space of the live studio audience, then the *affective stakes of affect's spill* become clear when Edward's work transgresses its inter-televisual references to instead place itself in the midst of a globally mediatised trauma referent. In the months following the September 11 attacks on the United States, Edward developed a new theme: a series based on 'channelling' September 11 victims. For viewers already attuned to the peculiar blend of TV formatting from which *Crossing Over* works, positioning the series' successful ratings recipe alongside a recent national tragedy was possibly the programme's next logical step, purporting to notch up its popularity even one more peg. Airing Edward's psychic communication with identifiable ghosts (as opposed to just any old ghost) offered to platform the new, dizzying heights to which reality-based entertainment could ascend – namely, the arena of national trauma, and indeed, a collectively experienced disaster. As word of Edward's programme circulated the broadcasting networks of the country, resistance mounted. In hours, Edward's episodes were banned from airing, deemed a coarse exploitation of September 11 for commercial gain. Edward has since tried to redeem himself from the media hunt following the filming of these live readings by explaining that they were vocationally necessary: 'how can I not deal with the biggest incident of death – especially when it's in my own backyard?'.[42]

While the filmed readings are not available for analysis (although some September 11 ghosts have emerged in non-themed readings from time to time), of interest in this account are the cultural contexts that produced both the necessity for Edward to 'read' September 11, and the subsequent banning of the themed episodes. If we understand psychic television as a forum by which memory cultures produce the logic and space for new practices of cultural memory, then we also understand that *Crossing Over* is a programme that performs a rationalising of what that memory is, or should be. In this sense, Edward's absented September 11 ghosts sit against other endorsed memorials to September 11 such as the World Trade Center site and the New York Fire Museum, which are understood to render justice to absent bodies, to formulate communities of mourning, and to reconsecrate the traumas of wounded space. These memorials, as Mary Marshall Clark explains, developed a practice of mourning tied to the national recovery of loss in the face of violence and to regaining autonomy for a citizenry placed into despair. They also, however, as Clark observes, mobilised a kind of urgency of reparation, serving the interests of United States foreign policy for which 'a dominant public narrative' of the event needed to be swiftly created. Marshall Clark suggests that the ability for civilians to articulate their own experience of the attacks in a variety of formats was interrupted by a rhetoric that produced an account of a 'nation unified in grief... allow[ing] government officials to claim that there is a public consensus... that has clear implications for national and foreign policy'.[43] It is surely this 'unification of grief' that Edward's themed episodes aimed to exploit, which is why their removal from the public eye is interesting: they could have ostensibly promised an enforcement of the exact 'dominant public narrative' to which other endorsed memory activities at that time contributed.

As media theorist Slavoj Zizek has argued, September 11 aimed to rupture how the Western world views itself through spectacles of the suffering other. Zizek describes a televisual economy that relies on divisions between First and Third World, and whose stability rests upon the hypervisibility of the Third World catastrophe: 'Somalis dying of hunger, raped Bosnian women, men with their throats cut.'[44] As Marshall Clark's oral histories point out, at the same time, overly familiar East–West scenarios of invader apocalypse meant that for many close to the event, September 11 was experienced as an after-event – an event that seemed to have happened in the image of its other Hollywood versions. The media's role in differentiating the privileged Western self from the suffering other has hence positioned the media as proxy progenitors of September 11, in that they are seen to be responsible for constructing the televisual space from which images of their own deaths are either omitted or become the 'saved' objects of cinematic fantasy. As Zizek would suggest, the victory for the perpetrators of September 11 has since been seen to rest not in the hijacking of airplanes, but in the hijacking of the Western world's scenic apparatus. In view of Zizek's position, one could argue that absenting *Crossing Over*'s September 11 ghosts from the framework of othering produced by the broader televisual sphere positions Edward's spectres as distinctly different from the interminably 'real' images the attacks aimed to bring into focus. In this sense, the restriction of these highly topical ghosts from appearing within TV programming incurs a 'meta' form of Edward's 'me-too' moment: it names an instance in which the ghost, as a secured and stabilised idea of appearance, begins to disappear. It is no longer a ghost; it is not even that. This, I suggest, positions Edward's televisual paradigm alongside the kinds of ethnocultural erasure that have been seen to have fuelled the September 11 attacks themselves.

As I explained earlier, me-toos reveal that there is an important 'structure of feeling' to be beheld in the logic of ghost construction, one which spectators use to rehearse their own affective reservoirs of body memory in order to obtain the social power that attends an object's 'feeling-causes'. The symbolic threat of a me-too reveals the larger cultural and political performativities of visibility that operate as part of televisual logics, and the ways that these performativities are made manifest through corporeal productions, circulations and precognitions of affect. These are performativities that manifest the way that the West claims itself and others its 'others'. The 'meta-me-too' caused by the missing episodes reveals the memory affect in the very moment that the *felt absence of feeling* appears, incurring a collapse, or implosion of the normative dimensions involved in everyday ways of feeling the memories of others. Its significance not only rests in how it reveals the building of an accepted ghost world, but also in how the occlusion of these highly topical ghosts goes to the heart of what is secreted by this programme more generally.

In the public moment of reckoning with the impermissibility of Edward's September 11 ghosts, a very different form of affective unfolding can be seen to sit underneath his formation of the ghost idea. This 'seething presence' hidden by the 'seething presences' that comprise *Crossing Over* forces us to recognise our complicity in the unequal economies of visibility that circulate public sentiments of loss. In this, the absence left by the missing episodes can be read as an affective spill that is unable to take heed of its own part in the act of destruction to which it consistently (unconsciously) refers. This spill *makes sensed* the ways that the memory affect promises to distinguish between remembering and remembered subjects at the same time as it contaminates those it hopes to individuate. Edward's missing spectres rather ghost the hypervisible sphere of Western hegemony itself, a scenic apparatus signified most lucidly in the image of a happy but dead family dog.

4
Affecting Indifference: Traumatic A-materiality in Second Life

Going digital

The Holocaust has gone digital. An interactive web tour of Poland's Auschwitz-Birkenau Memorial Museum; Anne Frank's secret annex online, which invites spectators to 'wander around' a virtual version of her original house and discover her 'secret annex'; and the Shoah Foundation's database 'IWitness', which promises mobile phone access to over 1,000 Holocaust survivor testimonies to enable 'critical multi-literacies for the 21st century... [such as] ethical remixing' are some of the recent examples of how Holocaust memory appears in the online era.[1] To date, digital forms of collective memory have been variously termed 'cyber commemoration', 'digital history' or 'cybermemorialisation'.[2] For the most part they have been judged in terms of their effectiveness as accurate representations of established histories. Other analyses alternately emphasise the forms of collective agency that are made possible with changed 'ecologies' of memory.[3] In this chapter I investigate how digital memories revision those material practices already established by the archives, memorials and museums prevalent in the latter half of the 20th century. I not only examine how digital practices transform memory texts, objects and sites, but importantly, how they repurpose the socio-cultural dynamics of memory that are held in the aesthetic frameworks they inherit. In this, I suggest that what they revision is the nascent materiality of the memory affect itself.

As I discussed in the introduction to this book, the First World canon of material memory culture is part of what Pierre Nora famously lamented as being responsible for the displacement of embodied memory realms with sites, and the creation of what Andreas Huyssen called the 'musealization' of the West.[4] Through their frequent emphasis on authenticity – the use of physical sites to present histories of trauma, the display of objects to present human loss, and the use of victim voices delivered in either live or video testimony – material memories have encouraged a primarily experiential relationship to history. They produce, as I argued in Chapter 1, a series of memory affects in which spectator bodies become gesturally, sensorially complicit in the production of ideas of pastness. I argued that canonical sites which represent the Holocaust not only contribute to the production of what Vivian Patraka calls the 'Holocaust performative', but also impact on the production of a Holocaust affect, which I defined as *the felt effect of the truth-effect* of meanings about its history.[5] I explained that when the Holocaust affect is mobilised, we embody the feeling of a particular past and we also embody the idea that this is a past that *ought to be re-felt*. I thus showed how the Holocaust affect came to designate the cultural circulation of the embodied capacity to secondarily experience an originary event within mass culture as a unique kind of 'first'. It also, importantly, designates the ways in which its staging as an 'unknowable' logic functions to create a distinct form of feeling (the feeling of an absence of feeling, perhaps), which in turn enables the production of varying kinds of spectatorial certainty.

In this chapter I bring the Holocaust affect into conversation with early discussants of online behaviour, such as Julian Dibbell and Theresa M. Senft, to contemplate how virtual practices make claims for the authentic nature of memorial artefacts, testimonials and material traces in an oddly mimetic synchronism with real-life counterpart memorials. Here I move away from the expansive museal memory industry to focus on the rapid electronic transformation of those very forms that have relied on material strategies of preservation. I examine the evolution of the Holocaust affect as it appears in one recent digital example, the United States Holocaust Memorial Museum's online memorial Kristallnacht in Second Life.[6] Reading this site as a remediation of the already mediated memory affects created by 'First Life' memorials discussed in Chapter 1, I argue that Kristallnacht in Second Life can be seen to be constitutive of a category of cultural activity termed 'virtual trauma'. Indeed, Kristallnacht in Second Life simulates the presence of a real-life museum that simulates an originary traumatic history. Thus, it reinforces those practices of feeling the memories of others that have been seen to generate forms of empathic witnessing, at the same time as it inverts key dimensions of those practices by positioning them in digital space. That is, if Holocaust memorials mediate history through material interactions with textual, narrative and artefactual remains, then Kristallnacht in Second Life remediates those mediations, such that it mnemonically refers to the very museal culture it also seeks to displace. To this end, I examine how Kristallnacht in Second Life constitutes new modes of spectator engagement, referentiality and their respective performative authenticities.

As I explained in the introduction to this book, aesthetics describes the mode of transmission between memory text and spectator as a constituting force, and I add here that with it, remediation accounts for the 'repurposing' of the dynamics of that originally 'material' exchange. I suggest that, just as virtual culture gives rise to different possibilities for spectators to feel or embody the histories of others, it also most often fails to interrogate the presumption of an ethics of reception produced by this form of exchange. To this extent, memory culture in 'Second Life' could be seen to participate in the continued circulation of a range of problematic 'wound culture' affects, such that it, along with the material memorial interfaces that have preceded it, serves the moral certitude of the spectator as witness over the subject being recollected. At the same time, the forms of affective engagement enabled by Second Life might also be understood as producing a kind of embodiment that enfolds existing practices of embodiment in newly embodiable contours. It does so through the way that it retrieves antecedent material frameworks for recollection and reframes them in new media formations. I consider this kind of concentricity in relation to Wai Chee Dimock's interpretation of remediation, which views it as a kind of genealogical *regenreing* that highlights the activity 'as cumulative reuse, an alluvial process, sedimentary as well as migratory'.[7] As Gillian Whitlock makes clear in relation to the digital evolution of forms of life narrative, 'Dimock's approach to the "reproductive" history of genres and media in terms of "kinship networks" suggests a methodology for tracking [its] transits...in terms of ongoing adaptation and appropriation.'[8]

By connecting virtual Kristallnacht to the material memory framework I have already charted, I evaluate the consequences of its evolution from physical to digital materialities for spectators, victims and survivors of collective trauma histories. As an extension of Dimock's insights, I further suggest that these consequences might be conceived in terms of not only recycled genres of recollection, but also of repurposed *affects* of recollection. I consider that, where in Chapter 1 the memory affect, operating as the Holocaust affect, was outlined as a specifically material eventuation of the interplay between physical sites and spectator bodies, here, the affective dimensions of material authenticity, which conventionally perform as a kind of unmediated source of sensation and sentiment, are made radically *a-material*. I hence open out questions around the transmutation of the memory affect as its sensorial networks become inverted, and I consider what becomes of a framework for feeling the return of memory when its primary sensory paradigms are subject to the kinds of undoing enabled by digital interfaces. In building on the concept of pre- and remediated memory affects outlined in Chapter 3, the memory affect here is remediation remediated, or 'hyper-remediation'. Below, I aim to understand the specific configurations that emerge from such practices of repurposed feeling.

Pixels in distress

As the historic incident of 'rape' in cyberspace made clear in the early 1990s, the online sphere is not devoid of trauma. The event came to notoriety after technology journalist Julian Dibbell published an article in *The Village Voice* contemplating the kinds of embodied subjectivity enabled by online communities, and the kinds of ethical and cultural repercussions that necessarily result when a broad spectrum of human 'real-life' behaviours are transported into the realm of the Internet. The 'Bungle Affair', as Dibble called it, raised questions about ontologies of the live as they connect to epistemologies of the virtual, about a somewhat disembodied sexual ethics, about the political formation of activist cyberspace communities and 'the curious notion of rape by voodoo doll'.[9] As Dibbell explained, the Bungle Affair occurred on one evening in LambdaMOO (one of the earliest online, multi-player, real-time virtual worlds) at a room in a venue used for its homely familiarity and its potential for playful interaction. Party-goers gathered, until the assailant Mr Bungle began...using his voodoo doll to force one of the room's occupants to sexually service him in a variety of more or less conventional ways...[I]n his private chambers somewhere on the mansion grounds [he]...continued the attacks without interruption since the voodoo doll worked just as well at a distance as in proximity.[10]

As the account continues, we understand that Mr Bungle's activities grew more violent: he began attacking other players – forcing them into unwanted liaisons, making them eat his and each others' pubic hair, and forcing one of them to self-harm with a piece of cutlery.[11] We understand that the after-effects of these acts – what the victim of this virtual rape *post factum* called 'posttraumatic tears'[12] – were also the embodied sensation of disembodied action: events that had been written in code and that had occurred in pixels had nonetheless generated a sense of corporeal violation. This doubling of virtual reality and real life was further complicated by the clumsiness of the technology itself, where 'raping' consisted of the un-permitted attribution of an action to a character by another player through the voodoo doll subprogram function – the description of one player's actions in text. In this way, as Dibbell put it, while some live bodies felt, '[n]o bodies touched. Whatever physical interaction occurred consisted of a mingling of electronic signals sent from sites spread out between New York City and Sydney, Australia.'[13]

While discussions around ontologies of the self as they are produced by virtual reality (or by new media technologies more generally) have come a long way since Dibbell's early account, as have the technologies themselves, the particular coincidence of the virtually traumatic as a highly specific – and equally proliferate – cultural paradigm of experience still remains to be fully explored. As Dibbell was right to foretell, his essay functions as a caveat for the more serious consideration of 'a future in which human life may find itself as tightly enveloped in digital environments as it is today in the architectural kind', such that, poised with the vista of an increasing cultural paradigm of virtual traumata, we might 'shut our ears momentarily to...techno-utopian ecstasies' and indeed consider the stakes of how the virtual sphere engages traumatic histories, enactments and workings.[14] For one thing, as Theresa Senft has pointed out, the potential for prosthetic identity-making practices offered by the convenient anonymities of cyberspace can often carry with it a kind of problematic rhetoric: the 'wrong assumption that only an online textual body is performative, whereas a biological body at the end of the terminal is stable'.[15] Senft makes it clear that the Internet ought to be considered 'a series of cooperative performance gestures from multiple computer and telephone systems', which also beckons us to ask, via Judith Butler, '[w]hich bodies come to matter – and why?'[16] As Senft explains, matter here importantly designates both 'materiality' and 'significance', and her reading of the Internet as a paradigm productive of bodies *that* matter, and of bodies *as* matter, opens out an important consideration of the kinds of spectator bodies constituted by the virtual reproduction of post-traumatic states and their histories.

Virtual trauma

Dibbell's account anticipates the range of cultural paradigms by which trauma and its historical legacies are now being virtually reconstituted. The new millennium's explosion in Web 2.0 user-interactive applications such as Twitter, Facebook and YouTube, as well as the increasing accessibility of cross-platform technologies such as the iPhone, has paralleled the democratisation and sensationalisation of the production of traumatic memory in contemporary life. Web 2.0 applications are typically distinguished from traditional electronic media for their creation of active 'produsers' of information, rather than passive consumers. While each new media platform generates specific processes for information dissemination, what they do have in common is their perceived potential for positioning the everyday citizen at the centre of meaning generation and its dispersal. This can be seen in instances such as the use of Twitter in the lead up to the 2009 Iranian elections, or in the highly mediatised death of Neda Agha-Soltan, which was posted on YouTube during those same elections.[17] For Anna Reading, the digital witnessing of the shooting of Agha-Soltan points to the formation of a 'globital' memory assemblage, in the sense that it is

> dynamic and involves transmedial, globalized, mobile connectivities and mobilizations. These traverse, reconfigure, and extend established memory binaries such as the organic and inorganic, the personal and the shared, the digital and the analog, the individual and the media organization, the local and the global.[18]

Here, the term 'globital' references digital memory practices as highly specific formations of power, in that it refers to, as Reading puts it, the 'synergetic combination of the social and political dynamic of globalization with digitization'.[19]

In broad terms, virtual trauma can be understood as the social, political and discursive dynamics of the remediation of traumatic events by, and in, contemporary digital culture. If we understand trauma via Cathy Caruth's reading of its inherent belatedness – the fact that it is the way that it is 'precisely *not known* in the first instance'[20] that haunts the traumatised subject – then rethinking trauma through, and *as*, the practices of 'globital' culture expands and reproduces trauma's own internal (and pre-digital) sense of virtuality. As Caruth herself explains:

> the experience of trauma, the fact of latency, would thus seem to consist, not in the forgetting of a reality that can hence never be fully known, but in an inherent latency within the experience itself... it is only in and through its inherent forgetting that it is first experienced at all.[21]

Given trauma's inherent latency, we understand that it is experienced as a kind of corporeal and psychic after-effect. Trauma inhabits the subject belatedly and as a searingly immediate form of belatedness: it is simultaneously dis/located, a/temporal and dis/embodied, or in Caruth's words: 'it is referential precisely to the extent that it is not fully perceived as it occurs'.[22] When the virtuality present in trauma's experiential condition becomes constituted by technologies of the virtual, which are themselves most often 'transmedial, globalized, mobile' formations of memory's material cultural conditioning, then we might also ask how the technologically virtual revisits the material sphere in which trauma's inherent structural virtuality is produced. That is, how might digital technologies restructure the incomplete perception situated at the heart of trauma's delay? Further, how might those technologies *'regenre'* not only history, but also the histories of representation that have carried that history? The digital reiteration of the memory affect might here be understood to recall unspeakability as a kind of aesthetic 'drag' in the sense that Avery Gordon evokes when contemplating those affect structures of 'the unhallowed dead of the modern project' which '*drag in* the pathos of their loss and the violence of the force that made them, their sheets and chains'.[23] The globital assemblages of virtual trauma drag in the pathos of its past traces as they were etched in the past. That is, the past can only ever return in Second Life complete with a remediated sensory imprint of its former mode of inscription.

Jay Bolter and Richard Grusin's definition of remediation as an activity that 'refashion[s] or rehabilitate[s] other media' hence becomes key to reading how culture generates virtual trauma.[24] Joanne Garde-Hansen *et al.* explain that 'implicit within remediation ... is always already a concept of memory: the memorialisation of an older medium by digital media'.[25] As Astrid Erll and Ann Rigney further explain, remediation is the force by which a '*dynamics*' of cultural memory might be activated: '[i]n this process, memorial media borrow from, incorporate, absorb, critique and refashion earlier memorial media. Virtually every site of memory can boast its genealogy of remediation, which is usually tied to the history of media evolution.'[26] Bolter and Grusin point out that remediation existed long before the digital era, in visual art practices, for example, but their thesis also emphasises what contemporary media bring to its functions: 'the transparent presentation of the real and the enjoyment of the opacity of media themselves'.[27] While this reading emphasises the mnemonic capacities of remediation – wherein the 'older medium' is memorialised as a techné by the newer one – Bolter and Grusin also point out that memory technés are not just vehicles for memory content that is held elsewhere, but themselves constitute meaning: 'because all mediations are both real and mediations of the real, remediation can also be understood as a process of reforming reality as well'.[28]

It is in this vein that Marc Redfield positions the rise of virtual trauma at the geopolitical turning-point of September 11 – a marker of global rupture, located trauma and the postmillennial realisation of the capacities of digital technology. Nonetheless, Redfield's account begins with linguistics: a discussion of the discursive performativities of nomenclature that frame the event now categorically known as '9/11'. In his argument, virtual trauma signifies the real of an event and its complicity within/as mediation: it is 'a making-legible, within the medium itself, of a violence inherent to all media technologies, which record and remember the unique only by effacing and forgetting it'.[29] The processes of effacement begin, however, in the forgetting enacted by such axiomatic phrases as '9/11' and 'Ground Zero'. The plural historical legacies these phrases might otherwise suggest (the 1973 Chilean coup in the case of 9/11, the bombing of Hiroshima in the case of Ground Zero), are erased by a linguistic deixis that subsumes all other reference points, erecting the United States event as a totality: 'you shall have no other September 11ths; should you mention others, they will be secondary to this absolute, toxic *punctum*:... to refer to Chile, you will have to speak of "the *other* September 11" '.[30] Redfield positions the performativity of the speech act 9/11 as a signature impulse of what he terms the 'vast representational and commemorative machine'. Its function as a name-date which forgets its own year enables it to provide the context for the mourning it would incite: it is 'a fetish precisely because it is always naming its own loss'.[31] This kind of duplicitous mourning, in which the term produces the effect it also names, is for Redfield idiomatic of a larger practice of virtual trauma, which generates dislocated, atemporal traumas through the spectacle of 'tele-vision', generating spectator-witnesses who identify with the experiential motif of 'a wound that... exceeds the difference between the real and the unreal'.[32] In this respect, it is the very enactment of the disaster as a kind of screen memory, or fiction, that deems its particular 'real' really real.

We understand via Redfield that the affective dimensions of cultural traumas are politicised, woven into what Ann Cvetkovich and Ann Pellegrini have termed 'archives of public sentiment',[33] or indeed, what are for Sara Ahmed the *'regimes of difference'* that are built through intersubjective practices of touch.[34] Such archives of public sentiment feed into what E. Ann Kaplan has called the 'empty empathies' of a trauma culture that spectacularly remediates itself over distance in an instant.[35] Mark Seltzer's study of serial killing in a cultural backdrop of 'wound culture' makes plain how addictions to visual spectacle come to generate new forms of public, considered as witnesses. He writes:

> the mass attraction to atrocity exhibitions, in the pathological public sphere, ... encodes, in turn, a breakdown in the distinction between the individual and the mass and between private and public registers. One discovers again and again the excitations in the opening of private and bodily and psychic interiors: the

exhibition and witnessing, the endlessly reproducible display, of wounded bodies and wounded minds in public.[36]

Implicit in these studies is a critique of the spectacularisation of trauma by the late capitalist culture machine. The seeming contradictions between archives as sentiments and empathy as empty, however, point to those collective formations of embodied memory that remember, as an internalised memory state, what might otherwise be construed as an always already surface impression. What is retained is the trace of a mediation that aims to produce the sensation of memory as new, as a first. That is, as iterations of memory's affect, virtual traumas imprint a trace or sensation of memory without its extenuating contexts of power. As Reading would argue, this comes to materialise a 'globital' dynamics of memory that is socially and politically contained in the performativities of digital aesthetics.

In my reading, a framework of virtual trauma understands that digital technologies remediate the mediating material frameworks that underscore the vast memory machine of the late capitalist West. This means that the digital realm not only produces new ideologies around how and what cultural memories should be, but also produces new perceptual processes *around those existing perceptual processes* through which its ideological frameworks then unfold. In this regard, virtual traumas are those cultural texts that have been produced by digital media in response to traumatic events, but are also importantly, the constitution of those events specifically *through*, and *as*, digitisation. Even more importantly, virtual traumas are the constitution of the event as originary *by* the particular performative functions of the technology itself, and doubly, are the after-effects (and 'after-affects'[37]) that these performative functions generate. In this, the digital produces a new formation of materiality out of the dematerialisation of originally material surfaces. Moreover, it produces new significances around the kinds of matter that do matter, along the lines of Senft's interpretation of Butler, wherein 'matter' exemplifies the intrinsic sociality of the unfolding corporeal sphere, here made digital. In this, the ways in which digital culture configures trauma become a resource for those very phantasmatic surfaces that affectively *make us*, such that, in the words of Cannon Schmitt, 'both media and mediation incarnate their own materiality'; they 'do not reflect or translate materiality so much as constitute it in their own right'.[38]

Feeling firstness

Richard Urban *et al.* point out that many of the conventions enabled by online worlds have led to the construction of a wide range of virtual museums. These combine the new forms of avatar experience enabled by digital technology with existing cultural practices for recollecting the past. In their survey, what becomes clear is that virtual museums can be highly experiential domains which

> allo[w] visitors to find out what it would be like to be caught in a tsunami (at NOAA's Meteroa Island), take a rocket ship ride into space (courtesy of the International Spaceflight Museum), or parachute from the top of the Eiffel Tower (in Paris 1900).[39]

They note, with possible irony: 'In a world where the sun always shines, there is no reason not to display artifacts in the open air or even floating in mid-air; since SL avatars are able to fly.'[40]

Given that artefacts can float mid-air and that visitors can experience a tsunami from the safety of a computer chair, the corporeal conditions by which the affective dimensions of trauma are communicated in Second Life are radically put into question. We not only need to ask about the rise of the memory affect as it is hailed into cyberspace, but also about its effects as a distinctly *repurposed* presence, that is, as a presence which *drags in* the originary fiction of its conditions as a specifically First Life form of emergence (which speaks to the originary fiction of a constituting, unspeakable trauma). If the Holocaust affect operated through those gestural regimes that make concrete subject positions a process of phenomenal certainty – certainties that are bound around the sensing of the unspeakable as an exemplary delimiter of intractable difference – then how might we understand the qualities of an experiential mode that both presupposes and repurposes this very dynamics? If the rhetorical affectivity of unspeakability can be found in its constitution as a limit point of recollection,[41] then what is the sensory condition by which we feel its virtual reiteration and how do we then also feel what it *drags in*?

While museums in Second Life represent a new frontier for practices of cultural recollection, they also perform a recall of cultural trends occurring long before the digital realm got hold of memory; they have the capacity to accelerate, hyperbolise or reject the experiential formations foregrounded in this book's aforementioned contexts of trauma tourism. In other words, they *re-affect the affective dimensions of the memory affect itself.* I discussed in detail in Chapter 1 the new ways in which relations between bodies and memory have been seen to incur a kind of 'prosthetics' brought about by the participatory strategies encouraged by new museum and memorial design. I also described the respective takes on prosthetic culture and affective forms of engagement put forward by theorists such as Alison Landsberg, Marianne Hirsch and Kaja Silverman, and how Landsberg in particular understands prosthetic suturing as a configuration of the mass circulation of image memories brought about by cinematic modernity. In response to Landsberg's thesis, I argued that understanding prosthetic memories as the automatically productive enablers of new forms of memory community neglected to point out the possible coercions by which the body's sensorial and emotional potentials work. Further, I suggested that prosthetic culture is deeply invested in the production of traumatic affect, the performativity of which is such that 'empathic' subjectivities are often produced as a result. In short, a critique of prosthetic culture underscores that the relational processes underpinning cultural memory aesthetics undermine the representational politics that those very aesthetics tend to performatively promise.

While for theorists like Patricia Ticineto Clough, affect precedes bodily motility and emotion – it is 'a substrate of potential bodily responses, often autonomic responses, in excess of consciousness' – it also operates on the level of 'self-feeling', in which the potential for relationships between bodies and other bodies, objects, or affects, is engaged.[42] As I explained earlier via Massumi, this kind of self-feeling can point to the intersubjective unfolding by which subjectivity occurs, such that '[s]ensation is the mode in which potential is present in the perceiving body' and it is 'a channeling of field-potential into local action'.[43] In Massumi's reading, in determining action and feeling, affect is the pre-individual state of potential that arises to eventually produce a sense of self. He offers this as a kind of 'intensity': '[t]he body...infolds *contexts*, it infolds volitions and cognitions that are nothing if not situated'.[44] We feel ourselves as relationally engaged in the kinaesthetic and spatial systems through which we become. For Clough, however, while there is a pre-social dimension to the body's experience of affect, it can also manipulate relations between bodies, and in this way denotes a sensing of sociality in process. She explains:

> Some bodies or bodily capacities are derogated, making their affectivity superexploitable or exhaustible unto death, while other bodies or body capacities collect the value produced through this derogation and exploitation.[45]

She further suggests that the sociability of affect is paramount in contexts of trauma, as is evident in 'the relationship made between victimized, terrorized, and hated bodies brought forth for the discourse and practices of counter/terrorism, surveillance and unending war'.[46] Mobilising affective intensity is then both an aesthetically ethical response to the project of 'witnessing' others, and constitutive of the terms by which those others come to be 'felt'.

In what Mark Hansen calls 'mixed reality' culture, the flat yet excessive aesthetic affectivity produced by virtual worlds can reveal the centrality of affect for understanding a more general technics of embodiment. That is, rather than understand a multi-user environment such as Second Life in representationalist terms, as that which simulates how First Life looks, Hansen argues that it needs to be understood in functionalist terms, as a simulation and extension of our primary perceptual processes. Hansen's notion of a body-in-code opens out the role of new media platforms in highlighting a fundamental technicity at the core of embodiment itself, a technicity mobilised and explained by affect and its transmission. As per Antonio Damasio and Thomas Csordas, whose respective arguments for a phenomenological unfolding of the self I discussed earlier in this book, Hansen suggests that a virtual cultural landscape might be understood to *'foregroun[d] the constitutive or ontological role of the body in giving birth to the world'*.[47] In Csordas' and Damasio's arguments, feeling is what mediates the production of the sense of self and its capacity to produce itself and others as objects such that, as Damasio explains, 'we only know that we feel an emotion when we sense that emotion is sensed as happening in our organism'.[48] Hansen proposes in similar terms that we understand mixed reality as a

> norm determining what perception is in the world today. Put another way, today's mixed reality paradigm makes ubiquitous

(specifically as a technical phenomenon) what we might think of as the experiential condition of mixed reality – that is, mixed reality as the condition for all real experience in the world today.[49]

While Hansen understands affect as pre-personal potential, along the lines of Clough and Massumi, he also differentiates this pre-individuality from interpretations such as Emily Apter's, which read it as the 'theory lite' take on 'virtual subjects who have little real stake in national, ethnic or gender affiliations, as well as the anticipation of a critique of posthuman subjectivity at the end of the millennium.'[50] For Apter, affect can only ever register what I called in this book the 'loss of loss' borne by late capitalist postmodernity, characterised in the flatness of ' "easy"... feelings washing about in a depoliticized space of the transnational commodity', as 'visual pleasure after aura has been drained off'.[51] While I have already discussed the way in which Jameson typified such easy feelings via postmodernity's 'waning of affect', it is important to note that Apter sees the wane itself as the defining affect of the contemporary commodifying 'imperium'. Distinct from Apter's critique of affect's apoliticality, Hansen finds the performative formation of 'virtual subjects who have little real stake in national, ethnic, or gender affiliations' acutely political for the very way in which the virtual experience of affect's wane can make clear that 'the only way to acquire an identity is to "pass," to perform or imitate a role, norm, or stereotype that is itself a cultural performance'. Moreover:

> By decoupling identity from any analogical relation to the visible body, online self-invention effectively places everyone in the position previously reserved for certain raced subjects: everyone must mime his or her identity.[52]

For Hansen, this experiencing of a self's performativity connects to its underlying pre-individual affectivity, and can thus make appear 'precisely the experience of one's incongruity with oneself (one's excess in relation to any fixing of identity)'. Affectivity, 'as the phenomenological correlate of singularity beyond identity' and 'that mode of bodily experience which mediates between the individual and the

Affecting Indifference 121

preindividual, the personal and the impersonal' opens out the very perceptual plane by which the sociality of such identity constructions is made material.[53]

For Hansen, it is the particular technicity of virtual reality that can re- and de-materialise the materially affective and affecting processes that constitute selfhood in everyday spheres. As he explains, the functional homology between virtual reality and the virtuality of affectivity is to be found in the way in which affect reveals 'a conceptualisation of the human as virtual in a very different sense – as always in excess of any particular determination'.[54] Given these perspectives on relations between affect and virtual worlds, the affective potential of a body performing as a wound, or alternately as witness to the wounded, might be understood to recur in Second Life as the Caruthian trauma in paradox. That is, the memory affect emerges as the curious iteration that repeats the fact of a latency that has not ever been known, *twice*. I suggest that this is a process that is enabled by, and constitutive of, a specific practice of feeling.

Kristallnacht in Second Life

Figure 4.1 Digitised photograph: Germans pass by the broken shop window of a Jewish-owned business, screenshot Kristallnacht in Second Life

Unlike the victims of Mr Bungle's virtual rape, I do not hold a close connection to my avatar. In Second Life speak, I am a 'newbie', and visibly so for the way my avatar's appearance is so seemingly generic. I look and feel too white, too perky and too young, dressed in an outfit I did not choose – a pink dress with small white polka dots, my head topped with a neat and purposeful pony tail. Even the way I am holding myself makes it clear that I am new around these parts. I am standing in a small museum newsroom, set up as a space in which I am invited to read a variety of information bulletins, before receiving a dossier with instructions that inform the investigatory mission ahead. The walls of the newsroom are covered with dominating black and white photographs that amplify a certain chronotope, denoting a field of sensation that I can only describe as historical. Wooden desks and books are placed around the room – one supporting a curiously antiquated telephone – but I find I cannot activate the phone. Instead I peruse the information displays, focussing on the captions illuminating their otherwise anonymous images of destruction:

> Caption 1. View of the destroyed interior of the Hechingen synagogue the day after Krisallnacht.
>
> Caption 2. Newly arrived prisoners, still in their civilian clothes, and after shaving and disinfection, stand at roll call in Buchenwald concentration camp shortly after Kristallnacht, 1938.
>
> Caption 3: Germans pass by the broken shop window of a Jewish-owned business that was destroyed during Kristallnacht. Berlin, Germany, November 10, 1938.

Eventually, instructions appear out of a glowing ledger on a desktop: 'As you leaf through the accounts of eyewitnesses, your mind takes you back to that night...and you begin to reconstruct the events of Kristallnacht through their stories.' While I slyly resent the use of the second person present tense to explain what I should be doing ('As you leaf...') and thinking ('your mind takes you back...'), I nonetheless follow through, stepping out of the newsroom and into the Second Life memorial titled 'Witnessing History: Kristallnacht, the November 1938 Pogroms'. As black and white pixels disintegrate around me, I arrive in streets of colour.

Embodying deferred deixis

Figure 4.2 Kristallnacht streetscape, screenshot Kristallnacht in Second Life

Kristallnacht in Second Life was designed by high school students in collaboration with the United States Holocaust Memorial Museum, with the idea that 'the folks that come through will not only learn more about history, but *absorb* it differently'.[55] The particular history in question occurred in Germany and Austria on 9 and 10 November 1938, when violent anti-Jewish pogroms facilitated the murder of almost 100 Jewish people and incarcerated another 25,000 to 30,000 in concentration camps. In the blogosphere, virtual Kristallnacht has been lauded as one of the most important contributions Second Life can make to real life for its approach to intergenerational remembrance. It can in this way be understood as a recent iteration of the Holocaust performative that performs with, and produces, the Holocaust affect. And yet, as I suggest, the particular kind of affectivity that it generates through its play with the Holocaust performative could itself be understood as a means to make the experience of affect re-experiential (or 're-affective'). Put differently, this is a landscape in which we experience the hyperbole of the memory affect as a distinct kind of phenomenal paradigm in its own right (materialised in not only the object of, but also the functioning of, the figure of the avatar).

As I explained in Chapter 1, the tourist's movement through a site such as Dachau Concentration Camp Memorial could be classed as a movement from 'in context' to 'in situ' frameworks of performing the past, where the 'poetics of detachment' more usually performed by ethnographic objects is expanded, so that the whole site functions as a material remnant indexing loss. I suggested that, particular to concentration camp memorials is the sense that the very 'in situ' veracity that they establish is built out of the history of loss that the *poetics of loss* seeks to mark. Tourists step into the loss marked by what remains at the site, and this is what constitutes the production of history's sensorial and emotive weight: their bodies are positioned as material reminders of what is precisely *not there*. They are at once the remembering mechanism and the material object that calls lost bodies to mind, forming a circular deictic affect flow between site, rememberer and history. While it imitates a site such as Dachau, Kristallnacht in Second Life is not a physical museum, nor is it a site-specific location. It instead holds a decidedly ambivalent form of indexicality to the history it wants to unravel for how it images, and inverts, a combination of First Life site-specific and curatorial design strategies. How the site then constitutes my avatar as a *sensed experience* constitutive of my lived material body becomes key to unpacking – as Senft has intimated – a politics and ethics by which some (memorial) bodies come to matter and why.

Like Dachau, the 'grounds' of virtual Kristallnacht aim to teach visitors factual history and then encourage them to imaginatively explore that history in a lived situational space. In this way it offers an investigative journey that is to be undergone by an avatar whose job it is to experientially reinterpret ruins by first factually informing the structure of feeling that they are to carry through those ruins. The entry point, the museum information centre, is housed in a larger building, which resembles a large bank, mausoleum or modern gallery. The décor of the newsroom resembles a partial reconstruction of a historical time period, offering a kind of ornamental semiotic to the instructional function of this space's images, texts and photo-captions, which prepare me for the experiential journey ahead. Here, I am given the task of interpreting history through curated artefacts displayed in a 'recognisable' exhibition space. These artefacts imitate the real, until one wall-sized photograph misbehaves: it dissolves, and I step through it to arrive on an abandoned streetscape portraying the aftermath of Kristallnacht. Testimonial words and voices haunt the streetscape, punctuating my movements with the grain of the disembodied 'real'. Photographs and documents have been digitised and positioned in abandoned buildings, re-performing the referentiality of the Barthesian 'this-has-been'. I visit a range of places, including one family's claustrophobic hiding space. I then travel to an atrium, the Reflection Space, in which I offer my own testimonial comments and see digitised video footage of the survivors who have contributed to this site in the final room, the Witness Rotunda (Figure 4.3).

Unlike Dachau, what Kristallnacht in Second Life produces is an interpellation of 'in situ' and 'in context' processes, such that the poetics of detachment that it inevitably expands upon is the object of museal culture itself. This occurs firstly in how those traditionally 'contextualising' practices, such as schemas of cataloguing, classification and display are made, in the first phase of the Museum's Newsroom, the focus of an 'in situ' approach by being enveloped in 'more of what was left behind, even if only in replica, after the object was excised from its physical, social and cultural settings'.[56] It occurs, secondly, in how that which follows the introductory Newsroom frame, the 'in situ' scenography of a deserted streetscape, is placed

Figure 4.3 Kristallnacht Witness Rotunda, screenshot Kristallnacht in Second Life

'in context' by its positioning in relation to the museum's first phase, that is, 'in relation to a classification or schematic arrangement of some kind, based on typologies of form or proposed historical relationships'.[57] In making the contextualising practices of curatorship themselves a form of 'in situ' landscape, and in placing 'in situ' practices 'in context', Kristallnacht in Second Life memorialises museal culture while also reproducing it. If what we experience at Dachau is a material remnant that refers, via our bodies, deictically to a past, then what we experience at Kristallnacht in Second Life is a pixellated referral, again via our bodies, to that original *process of* referral. In terms of what Kristallnacht in Second Life does to the memory affect, then, we might say that it uses affect to duplicate affect. Further, it is our bodies that are the mechanism that enables this process of de- and re-ferred deixis to occur.

In Chapter 1 I discussed via Jeff Pruchnic and Kim Lacey the processes of 'somatic marking' involved in the rhetorical uses of affect, in which present sensory experiences are informed by 'existing pre-conscious memories, particularly [the] recollection of images and marketing messages and the somatic markers created by those commercial appeals'.[58] In their work, Pruchnic and Lacey draw on Antonio Damasio's neuroscience, which finds, in ways similar to Casey's phenomenology of body memory, that affective responses are stored in 'dormant and implicit "dispositional memories" that record these responses within our nervous systems'.[59]

As Kristallnacht in Second Life is so heavily mimetic of First Life museal practices, it could be said to rely on the body's prior somatic markings to inform how the body, as a 'cooperative performance gesture' that works in tandem with the avatar, experiences the avatar *as a form of experience*. In this respect, the site's flat, planar landscape of pixels assembled around the semblance of material remains is generative of new memory formations to the extent that it first relies on those corporeal sensitivities that pre-exist it. At the same time, as a Holocaust performative that produces the Holocaust affect, it is constitutive of those prior knowledges as a corporeal necessity by which it can only then be read. What we experience of this highly 'globital' assemblage of the memory affect, then, is a form of experience that generates the necessity for an antecedent form of experience by which it can be understood, while also producing, via pixels, that very antecedence as fact; the site refers to a *prior-ness* that it simultaneously uses us to create. It is in this way that we not only '*absorb* it differently', but also, as I will argue, that what we absorb is notionally *experienced* as a form of affective *indifference* to the practitioning of embodiment itself.

The hyper-indexical aftermath

The dynamics of affective indifference can be seen most clearly in the site's choice to replicate the *aftermath* of the event rather than the Kristallnacht attacks themselves. Its deserted streetscape operates in the image of a traumascape, drawing on the tropes of what Maria Tumarkin has observed of the real-life locations of tragedies, which 'emerge as spaces, where events are experienced and re-experienced across time'.[60] While Tumarkin emphasises how the memory of place produces certain memorial affects, a traumascape might also identify a landscape that *performs itself* as experientially unique through an aesthetic crafting of remains, recognising that places *require* embodied action in order for the recollection of traumatic histories to occur. Traumascapes are culturally occasioned sites that are participated *within* for their meanings to be retained. And yet, unlike a site such as Dachau, this Second Life traumascape bears no material connection to a former historical trauma, and as such, has not 'witnessed' the memories of death that it evokes. It enfolds the avatar in a constant contradiction between the 'real' to which it would refer, and those conventions of behaviour that traditionally constitute it as a place of 'feeling'.

Lisa Saltzman's notion of the postindexical illuminates how the site works on this count. As I explained in Chapter 1, for Saltzman, the indexicality of a memorial artwork operates as a 'mode of making meaning in relation to the world that is predicated on physical contiguity, on material relation, on the trace of the touch'.[61] The postindexical, alternately, can be considered the 'empty index... the index that is no longer a sign, but instead, pure signifier' and is intrinsic to 'a renunciation of a certain relation to the real'.[62] Kristallnacht in Second Life is not postindexical, but rather *hyper*-indexical for its relentless amplification of a material memory formation to which it holds no physical contiguity. As such, there is no postindexical renunciation of the relation to the real but rather an obsessive, pixellated manufacturing of it.

While the streetscape poses a fictional materiality, there is a different kind of pixellated authenticity generated by the site's digitised photographs, which instead compose a converse kind of immaterial 'real'. The photographs are high resolution, and as digitised images of what perform as once-photographic stills, they punctuate the space's oversaturated cartoonness – making me look like a silly animation in the face of the reference they are able to produce. On this point, the final wall-sized depiction of Germans standing in front of a broken shop window renders a much clearer vision of humanity 'passing by' than my own polka-dotted state of ambulation seems to allow.

We understand through Marianne Hirsch that the photograph's autonomy as a memory object is gained through its survival as a material trace and its direct reference to what once was present.[63] Its material indexicality works to produce the perception of loss out of an absent presence in two simultaneous ways: it marks a survival into the now and the loss of those whom it documents. If analogue photographs index the real through two material contiguities to it (which also cross-reference each other), then a digitised photograph indexes *the index* of the real, producing an *augmented autonomy* composed of the object's interaction with a prior field of simulation. Their immaterial real throws conventional indexicality into doubt: the digital photographs in Kristallnacht in Second Life may not document anything other than the history of photography as a medium, for instance. While the digitised photographs rely on this process of ambivalent referral in order to be read as artefacts of history, they also undo the referentiality they hope to assemble by performing their augmented autonomy as a new mode of memory experience. 'My' ability to walk through a dissolving photograph draws attention to the hyper-indexicality underlying the entire site: the fact that this photograph might never have existed and possibly 'I' don't either.

If remediation is the refashioning of older media within a new medium, then the site's attention to those 'artifactual autonomies'[64] of conventional museum practice is what produces my own pixellation as a sensory process of affective indifference in itself.[65] In Kristallnacht in Second Life, the flat, planar vision of what I have described as the 'immaterial real' working alongside the 'fictional material' create the sensing of myself as an avatar who is undifferentiated from the pixellation that surrounds me. It is in its assemblage of the faked site-specificity of the experiential grounds as they rub against the artefactual remains of the photographic real that Kristallnacht in Second Life digitises a space in which I not only *encounter*, but *materialise through* an expanded, and oscillating, field of the memory affect. In materialising through this expanded field I am neither material nor immaterial and it is this very condition of a-materiality (in which I am neither not a body, nor *not not* a body) that constitutes the sensorial ontology of *affective indifference* which determines my overriding experience of this space. If I am affectively indifferent to my own ontology as an avatar, then the methods of mimetic duplication that this space assembles in relation to First Life memorial practices and culture, and that draw on processes of embodied unspeakability, become largely redundant. In this augmented sensorium, avatars and virtual museums can be seen, in Schmidt's words, 'not [to] reflect or translate materiality so much as constitute it in their own right'.[66]

My reading here is situated between Hansen's and Apter's divergent critiques of virtual culture to understand how the memory affect in Second Life operates as both a functional technics of the body and a descriptor for the sociality of that technics in process. In Hansen's model, virtual reality reveals the fundamentally virtual technicity of everyday human embodiment: it shows affect as a material process of relational self-ing. In this, the relationship between virtual reality and the virtuality of affectivity is to be found in the way in which affect reveals 'a conceptualisation of the human as virtual...as always in excess of any particular determination'.[67] In Apter's model, affect can only ever register the 'loss of loss' characterised in the flatness of ' "easy"...feelings' which come to inform its wane as a newly affecting genre of sociality in its own right.[68] While Apter is deeply critical, affect's wane is, for Hansen, acutely political in that it enables all forms of bodies to experience the same kind of performative (post)indexicality. In my reading, this performativity was only made apparent when the terms of my avatar's embodied experience imploded: it could fly, but couldn't walk though certain doors, for instance; it was given to perceive 'authentic' photographs, but could also *trespass* through them. In this, the sensing of my avatar's a-materiality as part of the clumsy process of negotiating this world was only ever an accidental means by which I felt myself as relationally aware. That is, the sensing of my avatar's ontology as homologous to my own everyday *being-ness* was a productive accident produced by the overriding hyper-indexicality drawn by the space, a space which more often offered to make me physically awash with an *augmented* sensorial paradigm of memory's affectivity in pure terms.

Here, Second Life, in 'mak[ing] ubiquitous... what we might think of as the experiential condition of mixed reality' enfolds that very tension between the 'mode of bodily experience which mediates between the individual and the preindividual, the personal and the impersonal'.[69] This tension became sensorially apparent only when the simulation of memory's affect accidentally catapulted me into a sensory perception of its undoing, wrought in the very space between the 'immaterial real' and the 'fictional material', to position my avatar-self as neither/and/or.

Re-affecting the wound point

Figure 4.4 Aryanised shop, screenshot Kristallnacht in Second Life

Inside the streetscape, there are anti-Semitic graffiti slogans written in German on building walls – translating as, for instance, 'You Jewish Pigs'. From a birds-eye view (my avatar can fly above the cityscape), the streets comprise a small town centre, which holds a synagogue at one end, a police station, some shops and a house with a hidden room. The space is designed for me to gradually uncover it – in many of the buildings there are digitised historical documents, photographs and archives of histories and events to encourage my active participation in reconstructing the history that occurred 'here'. There is also a repeated spoken transcript that echoes across the rooms and spaces, derived from the testimonial voices of victims of Kristallnacht who appear later in full-length digitised videos in the final annex. Sections of this text interrupt each other, to produce a background soundscape of disembodied voices:

> You have to tell the people what happens, so they know it will never happen again... It's terrible when you live under a dictatorship, it's horrible, it doesn't matter what dictatorship... We call ourselves homo sapiens, the wise man, but sometimes I wonder whether we should call ourselves... the most foolish one... If we stop this at the beginning... such horrors as the ghettos and the concentration camps would never happen.[70]

While Vivian Patraka's conception of the Holocaust performative points out that performance is accountable for its remediation of the 'goneness' incurred by an originary traumatic event, Senft has noted that the ontology of performance is challenged by online temporality: the Internet is a 'place that does not defeat death but is itself deathless'.[71] In terms of a Holocaust performative, then, the Internet doesn't do 'goneness' – rather it makes goneness gone. As a constellation of pixels, the deathlessness of my perky, bright little avatar body is made apparent in its ability to fly, to see across walls, and to bump into objects and corners without damage. My avatar does not experience a corporeal affinity with the historical specificities of traumatic loss; it does not materially feel and cannot materially 'die'. This is an ontological privilege that becomes particularly acute when my avatar is invited to engage with the histories of Kristallnacht victims.

One of the first activities my avatar occasions in this respect is the breaking of shopfront glass. The exhibit is titled 'Aryanized shop' – a shop called 'Strauss', which sells hats and luggage. Walking past the shopfront, my avatar activates the explosion of its glass window. As I do so, I shift from being 'journalist' to being a victim-perpetrator of the histories the site wants to portray. And yet, having flown into this space, and being able to 'walk through' other concrete objects such as desks and walls, the position from which I experience the smashing of glass is limited. Indeed, my own experience – or the site's presumption of my right to experience – seems indifferent to what I would be given to understand of the ontology of experience itself, allowing a similar kind of apathetic looking, or passing by, figured by the dissolving photographic image I stepped through in order to enter the streetscape described above.

In this, one could recall Redfield's notion of the 'violence inherent to all media technologies',[72] which in this case uses violence *violently* by effacing and erasing it *as* violence. It is the script that permits the perfunctory breaking of glass but that nonetheless produces me as an avatar engaged in the successful re-feeling of a history. If, as Sara Ahmed argues, feelings are 'produced as effects of circulation' and the ' "I" and the "we" are shaped by, and even take the shape of, contact with others',[73] then a certain problematic ethics of memory is to be observed in the way my avatar is given to feel, and not feel, the history of others in this place. This might be understood via the inverted sense that while the virtual re-enactment of violence in turn erases violence altogether, it nonetheless produces my avatar within the conceit of a spectatorial practice of prosthetic 'suturing', in which it is my avatar's a-material indifference that is given to empathically engage with the histories being portrayed. This is brought home with the knowledge that with the prosthetic extension given to me by, for instance, being able to fly, the very parameters by which I connect to any paradigm of 'experience' are undone. If I can fly, then photographs can be fake and histories fabricated. The functional homology here is what makes experience, as a paradigm of empathic and ethical knowing, problematic (Figure 4.5).

The redundancy of my avatar as a conduit of re-experience is made most clear in the site's attention to a primary, if hidden, wound point. In Chapter 1 I discussed how the Holocaust affect produces the central lacunae of trauma discourse, and with it, a self-congratulatory sensing of oneself in the process. Trauma discourse revolves around the very impossibility of perceiving the imperceptible, the sense that every narrative evolution holds a kernel of traumatic history that is not yet known, even to the narrator, but that nonetheless produces (and answers to) the testimonial fact of the narrative itself. This lacunae is enveloped in the spatial narrative of the Kristallnacht site: a hidden room within one digitally rendered family household. Unlike the other buildings planted across the streetscape, this building is

Figure 4.5 Hidden Room, screenshot Kristallnacht in Second Life

designed to render the distinctly personal over the institutional (police quarters, courtroom) or grandiosely sacrosanct (synagogue). The text that is spoken upon entering the room sounds: 'We don't know what is happening right now, but for now we will hide in the attic of our building...apples for the Fall...'[74] The site assembles me as a subject who has the right to explore what is presented as the privately personal. Its semiotics perform homeliness through the imaging of intimate spaces, cosy furniture and a bowl of red apples. It is claustrophobic inside, it is difficult to get out, there is nothing to do inside: I become its trespasser-cum-captive-family member.

In Mark Seltzer's terms, this room could be seen to attempt to manufacture for 'the pathological public sphere' an 'opening of private and bodily and psychic interiors',[75] which, in an inversion of his terms of study, fetishises not 'wounded bodies', but their absence. And yet the room instead draws into focus still something more, not absence, but its negation: non-absence, or the loss of loss. This can be understood in the way that the room inadvertently works outwards as a memory index, referencing those other, now 'globital' Holocaust assemblages and their attendant memory affects: the Anne Frank Museum in Amsterdam, famous for its Secret Annex (now perusable online), or my 'punchline' experience of the crematorium at

Dachau. As a fictionalised house representing the material history of other hidden rooms/houses, this house comes to perform as a house 'bereft' (of inhabitants) and a house bereft of 'houseness'. In this, it materialises virtual trauma and positions my avatar-self at the very centre of this appellative moment. As a virtual trauma, and as the centrally visualised blind spot within virtual Kristallnacht, the house operates as a fictional materialisation of a wound point to instead become 'a wound that... exceeds the difference between the real and the unreal'.[76] It also produces me as a spectator/saviour whose own a-material indifference exceeds (or possibly recedes from...) exactly what I am asked to perceive of this moment of 'encounter'. In the hidden room of a hidden house that never quite existed, I become the subject of genealogical *'regenreing'* that Second Life most carefully reproduces, as well as the process by which it occurs.

Ontologies of advantage

Senft's suggestion that discourses of virtual embodiment promote the 'wrong assumption that only an online textual body is performative, whereas a biological body at the end of the terminal is stable'[77] here reveals that the biological body at the terminal end is constituted by its relationship to its online self and online 'others'. As per Hansen's argument, this kind of intrinsic intersubjectivity is further generated by the mobilisation of trauma affects in online spaces as they functionally speak to the ways that we more ordinarily practice an everyday technics of the self. These are affects that produce us as differentiated in the very moments that we co-contaminate other bodies and selves. While the virtual traumas of Second Life can in some senses generate further binaries in discussions around memorial culture more broadly, it is not my intention to argue that Second Life memorials are 'fake' while First Life memorials are 'real'. Rather, I am interested in arguing that Second Life memorials – in mobilising the digitally virtual – highlight the contingencies, performativities and truth-effects of First Life memory culture. At the same time, Second Life memorial museums such as Kristallnacht participate in memory's wound culture affects in problematic ways. Mobilising affective intensity, as both emotional and sensorial potential – particularly as it relates to regimes of memory – here becomes questioned as an 'aesthetically' ethical response to the project of 'witnessing' others, while it is also constitutive of the terms by which those others come to be 'felt'.

Positioning the avatar's a-materiality as a form of experience that denotes the site's primary affect-experience is one which highlights the politicality enabled by those accidental unravellings of the memory affect. It can, in turn, enable a re-perception of the very perceptual processes that bring the self to being. I suggest that the accidental ontological rupture of feeling indifference, here incurred as an experience charted *between* the spaces of the 'immaterial real' and the 'fictional material', might be one kernel by which the site's most positive social force is produced. I consider, then, my own sensory awareness of this ontological *advantage* as the accident that gives way to perceptions of affect's movement at work. The 'advantage' staged through the avatar by the site unwittingly, unintentionally becomes the source of the site's undoing, which itself enables affect's spill to momentarily occur. When this constitutive frame of lost loss accidentally appears, I sense myself as a feeling being: I feel myself in the formation to becoming an individuated 'remembering body' at the same time as I am helplessly, hyper-indexically, affectively open. As a momentary undoing of the performativities of the memory affect, this exposes the experiential dimensions of memory's affectivity. In Hansen's words, it *'foregrounds the constitutive or ontological role of the body in giving birth to the world'*,[78] opening out the very perceptual plane by which the sociality of our everyday identity constructions are the virtual affect-effects of that which is always already material.

5
Affect's Spill: Theatrical 'Sensationship' in Cultures of Memory

Figure 5.1 *Tragedia Endogonidia BR.#04. Bruxelles*, photograph by Luca Del Pia

An archive of affects

Zürcher Theater Spektakel, September 2005. I stand with heart thumping at the closing images of Socìetas Raffaello Sanzio's *Tragedia Endogonidia Br. #04 Bruxelles*, and I am one of only several audience members remaining in an initially full house. The rest of the audience left this performance all at once, clumsily, loudly, in rebuke of having to see something that they had not come for: a scene which, it seems, had no place in the theatre at all. The scene was the relentless bashing of a man by performers wearing police uniforms wielding batons. The set housing these men was a vast marble cavern that, in the setup to the bashing, had allowed voluminous fake blood poured from plastic water bottles to swim across the white, opulent bureaucracy of the space. Before the blood was poured, the space had been cleaned. A black female cleaner entered with a mop: the rhythmic lethargy in her mopping occurring well before the entry of two men, who dressed themselves in police uniforms and set out crime scene markers as evidence and history of a death. They then poured the blood over a third man, who had meanwhile undressed to stand near-naked between the markers in preparation for his bloody, brutal beating. And then they beat him.

In the theatre of Socìetas Raffaello Sanzio, as Joe Kelleher and Nicholas Ridout have explained, the spectator 'is already in the archive, reexamining the state of the imprints'.[1] Innumerable scholars have since contemplated this scene to form a secondary archive of sorts, positioning it as an emblematic example of how a radical theatre aesthetics creates a postmodern ethics and politics, arresting as it does the spectator's embodied relationship to a certain 'fake' real established by its spectacular *mise-en-scène*. Maria Magdalena Schwaegermann, former Artistic Director of the Zürcher Theater Spektakel, explains that for her (and presumably for the audience of the 2005 festival) this scene became an endurance test 'to the bitter end.... so that the image [wa]s burnt into my brain'.[2] Helena Grehan recalls how she 'sat... squirming as the beatings continued',[3] and Joe Kelleher describes their 'interminable' sound as a 'nauseous crescendo mixed with interference' forcing the room to 'vomi[t] its memories at last',[4] to 'brea[k] through the mimesis'.[5] For Nicholas Ridout, the beating's staging was 'so obviously signalled... that it [wa]s experienced as something intolerable, intolerably real'.[6] For myself, the scene made me feel very much as if my cellular stitching, momentarily, might have just come undone.

Many such accounts testify to a theatre that feels the affect of its audience (the room itself is what 'vomits'), and doubly sears itself into the bodies of its spectators to enable – in director Romeo Castellucci's words – a somewhat violent 'penetrat[ing] [of] [the spectator's] shell'.[7] In doing so, they intimate that this distinctive form of corporeal 'penetration' evolves through a conscious complication of performance paradigms, or what Gabriella Novati describes as '[t]he negation of representation through representation and the negation of theatre through theatre'.[8] Such a process of negation might be considered via Rebecca Schneider's terminology, which argues in relation to re-enactment genres that theatre consists in both 'real *and* faux – action *and* representation', and it is this conundrum of the 'both/and' in which 'the represented bumps uncomfortably...against the affective, bodily, instrument of the real'.[9] Erika Fischer-Lichte, for example, argues that Castellucci's *Giulio Cesare* 'emphatically direct[s] the spectator's attention to the specific peculiarity and individuality of the actor's phenomenal body', resulting in a 'perceptional multistability' in which 'the order of perception...is upset...and another one has to be established'.[10] Bridget Escolme similarly observes that the same work's 'laying bare [of] the mechanics of representation insists upon an ethical relation...that "acting" cannot'.[11] Sam Trubridge suggests that *Tragedia Endogonidia* 'turn[s] speculation upon itself', 'excee[ding] the suspension of [dis]belief [sic] and hel[ping] to locate the audience's experience within a realm of material and physical reality'.[12] And Ehren Fordyce likewise argues that this inversion of spectatorship occurs through, and as, a theatricalisation of the 'border of semiosis' itself.[13]

The accumulation in the scholarly archive of these doubly affective stagings, where the world of theatre would firstly seem to perform our bodily affect and would secondly seem to affectively enter our bodies, suggests that when watching Soçìetas Raffaello Sanzio one feels a kind of *feeling* that ruptures how we as watchers are more usually used to being constituted by practices of vision. To be more specific, it seems that the very inversion of a mechanics of representation generates a particular dimension of affect through its process of theatrical self-negation. In this chapter I ask how theatre as a specific aesthetic object is able to help us re-perceive the more general affective dimensions at work in contemporary cultures of memory. These are dimensions that, as I have explained throughout this book, use re-feeling, and specifically the 'return' of memories as a process of generating perceptions of others, to draw spectator bodies into the ideological reproductions of contemporary memory regimes. In this, I have aimed to make clear that cultures of memory engage bodies in processes of sensing the past as a means to circulating that field of sensation as a truth-effect in itself: memory cultures produce specific forms of recollection that are highly affective and affecting. I here draw the questions raised in this book to a close by contemplating the ways in which the work of Socìetas Raffaello Sanzio draw the body's phenomenal perceptibility to the fore by seeming to *unstick* the spaces smoothed over by theatrical dissimulation. In connecting the stage mechanics by which Socìetas Raffaello Sanzio generate affect, to their theatre as a form of cultural memory, I suggest that they *de-remediate* the memory affect, and in so doing, enable memory's affectivity to be *meta-affectively* re-perceived.

Throughout this book I have charted moments in which the spill of memory's affectivity has been either consciously crafted through aesthetic forms and art practices, or as an eruption of a sticky and perplexing accident incurred by the stage or social drama's unravelling. At first glance, this affective spill might be considered in relation to Jill Bennett's reading of traumatic affect, which, operating as an autonomic corporeal realm, can be mobilised by particular art practices which aim to source a 'communicable language of sensation...with which to register something of the experience of traumatic memory'.[14] This, when 'properly conjured up, produces a real-time somatic experience, no longer framed as representation'.[15] In this reading (and Bennett is citing early 20th-century Swiss psychologist Édouard Clarapède), affect means that it is 'impossible to feel emotion as past...One cannot be a spectator of one's own feelings; one feels them, or one does not feel them'.[16] Art that helps us feel has the capacity to reduce cultural interference: feeling is 'not precoded'.[17] Bennett's reading of affect's intrinsic corporeal presencing leads to her more recent theorisation of the broader field of practical aesthetics, which understands the cultural role of art as one aesthetic milieu amongst others, and in this sense, as a means to underscoring a sensorial mode of attention that 'takes as its subject matter the already aesthetic nature of everyday perception'.[18] Aesthetics, as the medium and means of affect's transmission, produces and draws attention to our 'practical, real-world encounters'. Art thereby 'has the capacity to explore the nature of the event's perception or impression and hence to participate in its social and political configuration'.[19]

In the responses I've recounted above, the scholarship on Socìetas Raffaello Sanzio can be seen to register the spill of affect as something that is distinctly wordless, fleeting, phenomenal. It testifies to a 'squirming', 'interminable', 'intolerable' mechanism by which bodies become 'penetrated' and 'brains are burnt'. For my purposes, it is this forceful demonstration of the embodied nature of watching their work which also tells us something about the company's investment in theatre as a memory machine: that Socìetas Raffaello Sanzio are deeply concerned with enacting a form of cultural memory, or more specifically, that Socìetas Raffaello Sanzio *de-remediate* memory culture's investments in affect as the primary mediator of histories of loss. In this chapter I suggest that their works work to account for the spectator's body as a form of archival knowledge, understanding affect as a cultural and 'pre-cultural' form of bodily cognition; a corporeal archive of the sociality of 'public sentiments' that operate as a 'historical burden...on and for the body'.[20] Here, theatre is the aesthetic methodology that enables the re-perception of certain corporeally and affectively archived cultural traits. I suggest, in short, that through a vision of the spectator's body as archival, the theatre of Socìetas Raffaello Sanzio can be understood as an antidote to the sensorial, affective co-option of our bodies by those other culturally hegemonic practices of seeing and feeling memories that I have charted throughout this book.

In this chapter I focus solely on one scene within *BR.#04 Bruxelles* to suggest that Socìetas Raffaello Sanzio construct a meta-affective state for the spectator that is built out of a continuous collapse and rebuilding of relations between real and fake, sensation and spectacle. By meta-affective, I mean to indicate the tautologous nature of feeling, in that it occurs always, but often implicitly, as a kind of doubled activity. In the words of Jean-Luc Nancy:

> A subject *feels*... This means that he hears (himself), sees (himself), touches (himself), tastes (himself), and so on,...and thus always feels himself feeling a 'self' that escapes...or hides.[21]

Via Nancy, I suggest that the promise of theatrical meta-affect is the ontological rupture that enables feelings to be re-felt. These are feelings which, used normatively, produce ideals of subjective certainty when their processes most often undo those parameters of being to instead reveal to the self, the interconnection between its processes of self-feeling and self-being. As I described in Chapter 3, the ontological rupture that enables the constitutive frame of feeling to be re-felt occurs when we *feel* ourselves losing loss. I suggest that the work of Socìetas Raffaello Sanzio, in positioning us as phenomenally central to a process of feeling the loss of loss, creates the capacity by which *feeling can be felt to de-remediate itself*. As a momentary undoing of the performativities of the memory affect, affect's spill exposes the experiential dimensions of memory's affectivity, that is, it exposes the ways in which memory, as a cultural process of using feeling to make meaning, is causal: it figures 'our power to affect the world around us and our power to be affected by it'.[22]

A dramaturgy of meta-affect works to displace those practices that wed self to feeling and self to other, by tautologously *using feeling to expose its own processes*: it becomes central to the imagining of a politics and ethics of cultural memory. As Joe Kelleher has made clear, Socìetas Raffaello Sanzio theatrically assemble both the place of looking and 'the place where looking breaks down'.[23] Nicholas Ridout has contemplated this place via the historical conceptualisation of a Vibratorium, a mechanism that operates as 'an experimental apparatus for the exploration of intersubjective or social affect and its transmission'.[24] In simple terms, the place where looking breaks down might be understood as the place where Western sensory hierarchies begin to be inverted. More complexly, it might be understood as the place where the normative practice of social subjectivity is repurposed by a disturbance of the culturally unified relations between feeling, knowing and seeing. I want to contemplate why this rupture of the sensorium offers a fertile ground from which to conceptualise a meta-affective theatre of cultural memory, and why, in this sense, it hopes to undo those coercive operations of the memory affect as it moves *as* culture around and through us.

I firstly suggest that the beating scene within *BR.#04* articulates a theatrical response to central concerns faced by contemporary memory and trauma culture. In this, I argue that its form has much to offer, metonymically, post-traumatic postmodernity. I want to extend these discussions, however, by borrowing a phrase from Kelleher's own response to *BR.#04*, which observes 'a suffering that resides...in the image'.[25] The premise of a meta-affective register foregrounds how theatricality can articulate a conception of the remembering body in contemporary culture that moves it beyond simple alignments between spectator sensation of, and empathic identification with, a suffering other. An image that suffers places the burden of the wound neither in the domain of the object nor subject. It is, via Lisa Saltzman 'postindexical' in its suffering: it feels the feelings that we might otherwise be supposed to be feeling.[26] In this, *BR.#04* models a form of spectatorship that reconfigures how the body thinks it knows through feeling. It de-remediates the function of affect by refusing to use it to bind the suffering other to the 'empathically' engaged spectator. This form of spectatorship, as a radical kind of 'sensationship', forces us to *feel* feeling, and the truth-effects held therein. It hurts: morally, emotionally and physically.

Memory theatres

Jeanette Malkin's *Memory-Theater and Postmodern Drama* draws thematic and stylistic comparisons between postmodern drama and new millennial contexts of memory and trauma culture.[27] Malkin's research importantly assimilates modernity's seminal theorisations on memory through literature, philosophy and psychoanalysis (Proust, Benjamin, Freud) alongside modernity's signature historical events (Hiroshima, the Holocaust). She argues, via postmodern theorists such as Frederic Jameson and Pierre Nora, that we render ourselves culturally *through* these structures, that is, through '[t]he centrality of the fragment and the explosion...the "blasting open" of history and the collapse of linear and controlling "grand narrative"'.[28] While the epochal memory movements Malkin recounts point to specific memory topoi (deep memory, repressive memory, traumatic memory, amnesic memory), Malkin also implies that these specific cultural mnemonics coalesce to reflect a context that often integrates their multiple temporalities. We might then understand that, while prefaced by different disciplinary leanings, the traumatic belatedness so well theorised by Cathy Caruth, or Jameson's amnesic impulse, might not be mutually exclusive ways of experiencing subjectivity in postmodernity.

Malkin conceptualises postmodern drama's mnemonic impulses by analogising the architectonic memory theatre of Italian Renaissance rhetoritician Giuilio Camillo. As Malkin explains, Camillo's design for a seven-levelled wooden amphitheatre (which was incidentally never built) 'actualized the idea of "staging" memories by deploying a permanent set of images that were to provide a physical model for memorization', wherein 'a mélange of occult and mythic icons' were paraded before the subject.[29] The postmodern inklings of the structure are also apparent: 'Camillo's building was devised so that the spectator stood at its center, where the stage should have been, surrounded by the profusion of potent icons.'[30] As Frances Yates has explained, Camillo's model of theatrical architecture was designed to aid in the science of memory, that is, to improve what Yates terms 'artificial' as opposed to 'natural' memory via technés of good recall.[31] Its legacy, however, can be observed in how it theatricalises remembering, and more importantly, in how it positions the rememberer centre stage. Indeed, in this antique proposition for an architecture of remembering – one that is designed to aid relations between a subject and their memory – we nonetheless see a spatial configuration that already understands memory as somehow constructed, as somehow engaged in the operations of both performance and performativity. We also see a configuration that stages a body staging itself in the act of memory, making the rememberer the subject of the action.

While Camillo uses the theatrical metaphor to conceive of memory in spatial and visual terms, Malkin makes clear how postmodern theatre practitioners have used the memory metaphor to underpin theatre's relationship to the meta-traumatas of the 21st century. In this respect, Malkin characterises a theatre whose 'overabundance of disconnected stimuli: conflicting discourses, intruding images, overlapping voices, hallucinatory fragments' demonstrates a 'need to return and "rehearse" past moments or images, to repeat, quote, recycle... [offering] no aesthetically constructed release from the past'.[32] Memory theatre is that which both '*imitates* conflicted and sometimes erased memories of a shared past; and... that [which] *initiates* processes of remembrance through practices of repetition, conflation, regression'.[33] Her thesis preempts Hans-Thies Lehmann's own commentary on the structural mimeticism struck between postdramatic theatre and post-traumatic postmodernity, and particularly, his observations that the 'post' of postdramatic, like the postmodern, can be understood as an 'anamnesis' of drama.[34] What Malkin and Lehmann intimate is that the postdramatic, while deconstructive, is equally recollective, and thereby indicative of what Dominick LaCapra terms 'structural trauma': the loss evoked in the 'passage from nature to culture, the eruption of the pre-Oedipal or pre-symbolic in the symbolic, the entry into language'.[35] As a symptom of structural trauma and its possible remedy, the postdramatic pathologises *and* rhetoricises modernity's romance with loss.

In their roaming city-based repertoire *Tragedia Endogonidia* (of which *BR.#04 Bruxelles* is the fourth episode), Socìetas Raffaello Sanzio recollect structural *and* historical loss. Their memory theatre observes a kind of tautology of loss, which in its acts of incomplete recollection observes structural trauma, and in its thematic content, observes what LaCapra alternately terms 'historical trauma' – real-world events that have occurred, as in 'the Shoah or the dropping of the atom bomb'.[36] The *Tragedia* cycle contains the fading imprints of the West's fin de millennial atrocities, but this is importantly a form of tragedy that is already a remnant, a fragmentary formal order. As an endogonidial form, its creational self-replication positions tragedy as a kind of 'originary difference' in the Derridean sense – as that which, in longing for wholeness, only ever discovers its own multiplicity or rupture. Romeo Castellucci explains this contradictory impulse by understanding tragedy as a mechanism which presupposes death – 'an end (of the hero)' – and the endogonidial as alternately reproductive 'ad infinitum'; a mechanism's 'division of itself by itself, a sort of fall-out of spores'.[37] Their dramaturgy thus politicises what it is to recollect loss through what are themselves lost forms – particularly those forms, such as tragedy, which hold a specifically recollective function. *Tragedia Endogonidia* mourns the sanctity of tragedy as a formal code, producing the spectator within a logic that inverts the European theatrical sensorium as well as the symbologies of the late capitalist West.

In this dramaturgical logic, heroes are fleetingly witnessed through gesture, narratives become whittled down to sound bytes, and the chorus is evacuated. *B.#03 Berlin* offers an homage to the 'mute, tragic stature' of Germany's revolutionary Ulrike Meinhof with the vista of a mourning mother masturbating desperately on the edge of her bed.[38] In *A.#02 Avignon*, a stage direction reads: 'the gold room is covered with white fabric... On the right of the stage, three epileptics'. It continues: 'After a period of general convulsions two of the epileptics suddenly stop, then get up and walk off.'[39] The logic of such performance sits in how it balances readability against inexplicability – a convulsing body is extracted from teleological narrative but works viscerally to shake an audience into new kinds of perception. It is important to recognise that while the company aims to access a 'pre-ontological' state through pre-tragic form,[40] the equally implied 'post' of their engagement with attic tragedy registers a postmodern ambivalence for how historical traumas can be unwedded from the image apparatus around them. For this reason, scholars have varyingly remarked on what the 'post/pre' status of their response generates: a kind of pre-ontological affect summonses what Ridout has termed the 'material, feminine, infant and... animal',[41] and what Freddy Decreus has described as 'that point beyond language, beyond that first system that differentiated meanings'.[42] At the same time, any recovery of origins only ever pinpoints originary difference: the entry of tragedy as the mark of the law, that 'constitutive element in the establishment of a certain political order: that of the city and the class struggle'.[43]

The virtual traumas of the 20th-century West here become responsible for the failure of tragedy to redeem loss from the spectacular image sphere in which it occurs. This is a vision of tragedy that understands that the 'postpolitical, postdemocratic and post-tragic'[44] aspects of the age mean that even feeling has been conquered by commodity. Romeo Castellucci explains that tragedy becomes 'a mechanism to expose the dead body',[45] and Claudia Castellucci further suggests that theatre can enable a politicised apprehension of the dead body that goes beyond the consensus-making performativities of the media spectacle. This arises in how an 'exposure of the "reality of representation"' can force a 'fall into representation', which destabilises conventional spectatorship and enables other modes of perceptivity to emerge.[46] A vista in which tragedy has been banalised by its own perpetuating mediality is what drives this conceit:

> Disasters and the slaughters of innocents are everywhere referred to as 'tragedies', but this is an idea of tragedy that does not know how to distinguish these things from spectacle; nor how to think of them in terms of political crisis; nor how to gather them up on behalf of a metropolitan community, among people who are at the same time amassed and dispersed.[47]

In this regard, the company recollects how structural forms such as tragedy are implicated in histories of violence and yet fail to restore these histories to contemporary recollection. In this post/pre tragedy, the tautology of tragedy's own denouement sits squarely in the question of how the loss of tragedy might itself be restaged. The significance of the company's practice is in their capacity to strike a *re-perception* of this *loss of loss* and enable it to be provisionally found.

In positioning the figural dead body at the centre of their work, Socìetas Raffaello Sanzio produce a theatrical form that answers to the repetitions of traumatic scenography that haunt the mediatised stages of the West. They further intimate an important metonymy between the figural dead body to be revealed by tragedy, and that other figure who, according to Camillo, stands at the very centre of the theatre, performing remembering. This envisaging of the spectatorial-cum-remembering body as central to the mnemonic field of culture suggests a kind of synaesthetic, or affective, grounding of memory work that operates through a body's sensory, corporeal, even cellular systems. While in Castellucci's theatre, spectators are made witness to (or indeed refuse to become witness to) the scene of a body's brutal disappearance, in Camillo's memory theatre, the rememberer stands centre stage, engaged in a kind of affective contagion with the textual traces stored around her: a figure who, from the perspective of hindsight, we now re-encounter.

Un-feeling feeling

In *BR.#04 Bruxelles*, the brutally loud thump of the beating action was technically enabled by the real-time amplification of rubber batons as they charged against the performer/victim's body. Because real-time sound was amplified (microphones were implanted in the policeman's truncheons), the audience was not subjected to hearing a sound that merely sounded like beating; rather they heard the sound of beating being made. And as a variety of sources indicate, it seems that this very durational act of beating was what possibly forced an early escape for many spectators, as if they were themselves being beaten, as if their bodies resisted – not the level of the noise *per se* – but rather, its capacity for corporeal penetration.

Scott Gibbons is the long-time collaborating sound artist with Socìetas Raffaello Sanzio. His repertoire has been widely lauded for its unique techniques of 'micromusic' and 'isolationist ambiance', in which he manipulates one particular sonic register (water or stones) on what has been described as a 'molecular level'. In the company's words, this means: 'inserting potentially unlimited chromatic variants into normal notation in such a way as to give each sound a plurality of timings and durations which have nothing to do with unifying metric form'.[48] In the *Cryonic Chants* series, for instance, sound takes central focus and becomes a kind of material register that supersedes the ritualised gothic bodily happening occurring in the space.[49] As Daniel Sack has noted, Gibbons' score for the *Tragedia* cycle was celebrated for being 'based entirely on the human voice digitally manipulated into ear-splitting screams and groans that approach the limits of audibility'.[50] While molecularisation of most of the sound in *BR.#04* was achieved by using remastered recordings taken from an audio-file of the body's noise, the very 'seed of the sound' in the bashing sequence was created in the shared-time space of the audience, and was composed, 'molecularly', of the act of rubber touching live flesh. In this respect, when we understand that the 'composer's decision to use only voice-based sounds makes his music part of the truth, a sign of the truth', the 'sign of the truth' in the bashing sequence was rather the actual generation of this kind of sonic authenticity.[51] Indeed, the performer's body and the amplified truncheons were required to make the real-time sound that illustrated the scene it was occasioned by.

In Lehmann's argument, postdramatic theatres offer 'a transition from *represented pain* to *pain experienced in representation*'.[52] As I recounted above, the beating scene was experienced as 'interminable', a 'nauseous crescendo', a cause for 'squirming'. And yet, the affective spectatorial response and the refusal to watch that I witnessed occurred *despite* the fact that the meta-theatrical markers positioning it within the 'faux' field of representation were clearly laid plain: there was fake blood poured from plastic water bottles, the performers dressed, undressed and redressed themselves in costumes, and so on. In this respect, the particular spectatorial experience of *BR#.04 Bruxelles* occurred between these two opposing parameters. Indeed, the transformation of perception incurred by the work emerges from the interpolation of the twin paradigms of theatre and performance, or 'fake' and 'real', and more precisely, the *felt effect* of their coincidental enactment as it incurs (and undoes) memory's affect (Figure 5.2).

While Ridout has observed that it is the 'tendency to see the "real" ' in the company's work which 'is in fact an effect of the success

Figure 5.2 *Tragedia Endogonidia BR.#04. Bruxelles*, photograph by Luca Del Pia

of their theatrical pretending',[53] the inverse is simultaneously made to occur in this scene. In this, the success of their *realness* is what reveals simulation. Or, put differently, the ability to see the 'fake' in their work is an effect of the success of their non-matrixed enactment in real time. While the former tracks the representation of pain such that it might generate an empathic spectatorship via the success of its pretence, the latter might be seen to counter-track the representation of pain to instead reveal the pain in, *and of*, representation itself (to intentionally misread Lehmann). Susan Sontag might understand this counter-track as revealing a certain 'iconography of suffering' contained by news media, an iconography that has nonetheless become part of the 'quintessential modern experience'.[54] In Sontag's meditation on the spectacle of atrocity, what is observed is the capacity for the media to flatten an event, such that 'a catastrophe that is experienced will often seem eerily like its representation', and to generate relations of power through its remediation: 'a suffering that is outrageous, unjust, and should be repaired' only confirms that 'this is the sort of thing which happens in that place'.[55] It is these kinds of remediations that are, in different ways, tracked across chapters 1 to 4 of this book.

Sontag's reading of the mediatised iconography of suffering runs parallel to Kelleher's aforementioned reading of *BR.#04* as containing a 'suffering that resides in the image'. As I explained earlier, an 'image that suffers' places the burden of the wound neither in the domain of the object nor subject, indicating a kind of postindexicality to the cultural representation of histories of trauma and their discursive re-renderings. In *BR.#04*, its depiction of suffering (the moment of beating) is dramaturgically dissected while it continues its sonic penetration of spectator bodies. This dramaturgical dissection literally pulls the image apart even though its affective impact continues to work sensorially in the space. In forming an image that suffers, the subject in the image is continuously disrupted from ever completely entering the domain of representation; they cannot completely become the realm that is 'the mediatized iconography of suffering'. Instead, the subject in the image – the body that is being beaten – is sustained outside of the very image sphere that their body has been nonetheless appropriated to produce.

Here, what we instead witness is the body in the process of becoming an iconography of suffering, and more importantly, we witness our own affective relation to the split dimensions of visual upset and sonic penetration that this singular body is able to charge. In this, *BR.#04* models a form of spectatorship which reconfigures how the spectator body thinks it knows through feeling: the image itself is what feels the feelings that we might otherwise be supposed to be feeling. These tautologous feelings, as they connect to a particular and located historical trauma, and as this particular trauma connects to a cultural practice in which trauma is publicly displayed in the name of 'cultural memory', rupture how it is that citizen bodies come to respond to, feel and circulate the lived experiences of others in the name of empathy. Pain, as indistinct from the spectacle in which it is remediated, instead culturally presents the loss of loss itself, or to paraphrase Romeo Castellucci: presents an idea of tragedy that does not know how to distinguish itself from spectacle. In the beating moment, we are able to enact a politicised apprehension of the dead body, because we re-perceive the political machinations that have spectacularly created it. This is a moment that, as per my discussion of Ridout in the introduction to this book 'confront[s] its spectators or participants with something radically other, something that could not be assimilated by their existing understanding of the ethical...issu[ing] a demand they did not know how to answer'.[56]

It is the simultaneity of both tracks – represented pain and pain in, and of, representation – that cannot ever reveal the bashing of a body from outside of the mediated sphere in which it occurs. Doubly, it is the simultaneity of both tracks that creates an affective, embodied relation to the bashing, and at once disables the parameters of this spectatorial encounter. What can be revealed, and countered, is the spectator's more usually unproblematised 'felt' relationship to the spectacle itself. In this sense, I am at once feeling with another in their feeling of pain, and I am also feeling my own seeing of pain; I watch the durational nature of this act of pretence. In my reading, it is the conjunctive 'both/and' of these dramaturgies that accesses the bodily archive of the spectator, and arrests how that archive has been formed in – and as – a process of politicised affectivity that we come to call 'cultural memory'. The transformation of perception enabled here works to reconfigure how the body thinks it knows through feeling, such that feeling becomes un-felt, or twice felt. In this, it de-remediates the function and properties of affect, by refusing to use its enmeshment with a spectacular media/memory apparatus to bind the staged, beaten body on stage to the 'empathically' engaged spectator. It is precisely this duality that causes affect to spill, and for that spillage to create a form of spectatorship that hurts.

The empathy affect

As I have discussed elsewhere in this book, Elaine Scarry's prompts for thinking through the processes of pain suggest that when in pain, there are never enough words to typify that pain exactly. When not in pain, it is impossible to 'feel' the pain of another, because true feeling would necessarily involve being in the same kind of pain oneself. This is what Scarry calls pain's 'unsharability'.[57] Throughout this book I have examined the ways in which the unsharability of embodied feeling, alongside many other 'uns' spoken by trauma discourse,[58] has been placed in tension with the alternate prefix 're', thanks to a number of perspectives put forward in performance studies.[59] I have also discussed how performance studies theorises how bodies can get close to knowing and feeling each other through re-feeling – as a kind of choreographic history – the coalescence between embodied action (performance) and its signification (performativity).⁶⁰ Within these forms of discussion, the practice of feeling *as* another body (feeling as kinaesthetic movement, and thereby as motorsensory empathy) is most usually invested in an ethics of feeling *for*, or about, another body. As Susan Leigh Foster writes: 'like "kinesthesia"', the origins of empathy 'nam[e] the experience of merging with the object of one's contemplation', as an 'experience undertaken by one's entire subjectivity'.⁶¹

Theorising the unsharability of feeling in models of spectatorship, however, relies less on arguments for forms of embodied repetition on behalf of spectators and more on ideas of an affective 'arrest' of the viewer. Affect here extends ideas around physical sensation to include the physiological propensity for the transmission of bodily intensities: 'those resonances that circulate about, between, and sometimes stick to bodies and worlds, *and* in the very passages or variations between these intensities and resonances themselves'.⁶² Affect comes to explain how feeling is transmitted, shared, felt. For Denise Varney, the form of spectatorial experience established by 'postsemiotic' theatre works such as *BR.#04* can be understood via Brian Massumi, who posits that in a highly mediated, 'post-ideological' era, the power of an image resides in its felt 'intensity', such that 'semantically or semiotically-ordered levels of analysis are no longer adequate to the task'.⁶³ In Varney's terms, 'affect theorizes responses to the arts that are not only cognitive, but also...sensual and erotic as in our unmediated, "autonomic" or involuntary responses'.⁶⁴ In both readings, affect is the mechanism of a highly mediatised culture, and the possible aesthetic antidote to its effects.

While much has been made of the structural relationships between postmodern theatre and memory, less has been made of how its mechanisms for affectivity connect with a broader politics or ethics of cultural memory. In the models of spectatorship offered by Varney and Bennett, the sensuous complicity of the spectator with their mediated world is remediated by an arts practice that speaks in corporeal, rather than purely cognitive measures. Further, for Bennett, affect has a role in enabling a form of sensorial empathy towards the subject of trauma that works against the kinds of 'crude empathy' that merely offer 'feeling for another based on the assimilation of the other's experience to the self'.⁶⁵ In Bennett's model, empathy is 'grounded not in affinity (*feeling for* another insofar as we can imagine *being* that other) but on a *feeling for* another that entails an encounter with something irreducible and different, often inaccessible'.[66] Empathy here is deeply aligned with an argument for affect's effectivity: indeed, empathy comes to justify a range of artistic responses within memory and trauma discourse as a mootpoint practice of a range of ethical spectatorships. And yet, arguments for an empathic politics and aesthetics of relations tend to overlook the gritty particularity of this affect, and particularly, Sara Ahmed's observations around its impossibility, where empathy 'remains a "wish feeling", in which subjects "feel" something other than what another feels in the very moment of imagining they could feel what another feels'. This occurs such that 'even when we feel we have the same feeling, we don't necessarily have the same relationship to the feeling', intimating that at the heart of empathy there exists an important, but often overlooked, discordance between an intention to enact it and the actuality of performing it.[67] Quite cynically – but interestingly – perhaps, empathy might be one of those affects which diverts itself from its own cause.

While it might not be possible to feel emotion as past, it might be possible to feel emotion as doubled, or even tripled or quadrupled. In this respect, we might then ask how Bennett's argument for empathic, affective engagement sits against a contention such as Ahmed's – that empathy only ever operates in the future tense, as a wish, as deceptively multiple or even perhaps *dissimulating*. In view of the kind of spectator response I witnessed at the 2005 staging of *BR.#04 Bruxelles* we might further ask whether there is a form of dramaturgy that might arrest the *empathy effect* of *affect* itself. That is, is there a stage mechanics that can reveal the affectivity of empathy as it moves through bodies as *a* felt dynamics of cultural memory, as that which creates social effects and affects?

Spectatorship that hurts

In his philosophy on the materiality of sound, Jean-Luc Nancy qualifies the fundamentally meta-affective state of all sensation as well as the sensory means by which bodies repress this:

> A subject *feels*... This means that he hears (himself), sees (himself), touches (himself), tastes (himself), and so on,... and thus always

feels himself feeling a 'self' that escapes... or hides... as long as it resounds elsewhere as it does in itself, in a world and in the other.[68]

While Nancy gives credit to Aristotle for his thinking, and indeed, there is nothing particularly novel about the notion that sensation is implicated in how we know and make ourselves, there is possibly something helpful in the proposition that we self-reflexively 'feel' ourselves 'feeling a "self"'. This implies that subjectivity is a necessary engagement with the tautology of sensation, and that it nonetheless suppresses this tautology in aid of producing a seemingly unified practice of sensory perception, such that 'I feel myself feeling' becomes downgraded to 'I feel', and so presents the self-possessing unified subject to, and as, and of, and in, the world.

Nancy's focus on the tautology of sensory perception connects to similar observations on the tautology of affect. Within this vision of a kind of amorphous and multidirectional transmissibility, meta-affect becomes a *felt* awareness of the transmission – or transmissibility – of other embodied feeling states that occur within and between 'co-constituting' bodies. Meta-affect observes that feeling necessarily occurs in duplicate and triplicate, and – *via feeling* – makes feelings apparent to each other. In the field of anthropology, for instance, meta-affect has been used to describe how emotions become socially intelligible. Greg Urban observes in Amerindian Brazilian ritual wailing 'how one emotion (sadness) points to or "comments upon" another emotion (the desire for social acceptance)'.[69] In this, the ritual mourns the deceased and secures acceptability for the mourner in the social order. Similarly, Helga Kotthoff observes in Georgian ritual lamentation various forms of 'feeling rules'[70] that connect the dominant affect of grief to other cultural values of 'communality, gender, regional identity, morality, religion, social hierarchy'.[71] For Kotthoff, the 'aesthetics of pain'[72] within the performance actually encases the possibility for a broader 'reproduction of culture' itself, such that affect networks of gender, family and agency are themselves contained within the ritual's gender-based norms and codes.[73]

In his discussion, Nancy is particularly concerned with how listening is differentiated from hearing, arguing that 'to listen is to be straining toward a possible meaning, and consequently one that is not immediately accessible', while sound is that which is known and heard.[74] Nancy further makes clear how the act of listening works in relation to the optical register:

> In terms of the gaze, the subject is referred back to itself as object. In terms of listening, it is, in a way, to itself that the subject refers or refers back... In still other words, the visual is tendentially mimetic, and the sonorous tendentially methexic (that is, having to do with participation, sharing, or contagion).[75]

In this reading, the sound that is occasioned through one singular body (the body being beaten) resonates through the spectatorial bodies of the audience such that the 'sign of the truth' becomes a kind of uncontrollable contagion. What is being spread, however, is not the spectacle of a beating, but the very 'molecular' sonic components that are transmitted by it. The contagion brought forth by the materiality of molecular sound as it enters the spectator's auditory system and circulates through the cochlear receptors produces sound as material, and as a fleshly signifier that enters the fleshly gap of the body's ear drum. In terms of Erin Manning's reading of the politics of touch, it might be seen to sound itself through a body, relationally engendering a process of tactile, sensate individuation. In Manning's words, this involves the 'proposition... that touch – every act of reaching forward – enables the creation of worlds. This production is relational. I reach out to touch you in order to invent a relation that will, in turn, invent me.'[76]

But how does this moment of affective listening connect, firstly, to the act of spectatorship being undertaken, and the scene being witnessed, and secondly, to a methodology of spectatorship as respondent to the concerns of broader memory and trauma culture? What is brought about by a theatre that enables the memory affect, and specifically, its empathy effects to be meta-affectively revealed?

While the audience witnessed a clearly self-reflexive representation of a bashing, occurring within this was the necessity to sonically *produce* (not merely *illustrate*) the scene with the non-representational, durational *act* of bashing. In this respect, the sound molecules are what entered the fleshly molecules of spectators in the space. As an eruptive event, the amplified violence manifested within the skin, eyes and ears of those watching. The affective contagion here – the simultaneous *becomingness* of beaten body and spectators – occurs, if we agree with Nancy, as a methexic practice of self-extension or self-sameness on the level of sound, even as we simultaneously know, on the level of vision, that the pain being witnessed is an enactment. In hearing (becoming, even) the sound of a beaten body while seeing the correlative act of theatricalising beating, my own sensorial subjectivity is suddenly made plain: I hear myself hearing, see myself seeing, feel myself feeling, as my processes of hearing, feeling and seeing synchronically *upset* each other. What happens here is a startling process by which, through affect, I simultaneously perceive affect as well as its affect-effects, at work. That is, I feel feelings at work as they promise to make remembering and remembered subjects culturally distinct at the same time as they contaminate the very bodies they hope to individuate. I feel exactly those processes by which feelings possess us, such that we feel ontologically unique, separate and self-bound in the very moment that we might just be radically, affectively open.

The transformation of perception enabled by this scene creates its own meta-affective glare, in which affects upon affects seem to touch, undo and redo each other. In this, the empathy effect cannot be produced to merely bind me in a relation of power to the staged, beaten body, as in Sontag's sense of a 'suffering that is outrageous' but 'the sort of thing which happens in that place'. Instead, I feel myself feeling about, through and for the pain of another. This is spectatorship that truly hurts, and indeed, is what might just cause a rushed, hurried exit from the theatre space. If feelings are the archived bodily trace of an affect, then Socìetas Raffaello Sanzio reconfigure how the body thinks it knows through feeling to create a meta-affective theatrical politics and ethics of cultural memory. Here, affect spills in exactly the right way.

Affecting de-remediation

Bennett's conception of aesthetics as a means of re-apprehending those perceptual processes that form the cultural paradigms that enclose us in memory cultures can, in relation to the sites charted throughout this book, highlight how, where and why the memory affect is put to cultural work. In doing so, an aesthetics of meta-affect best reveals how memory culture works to produce the memory affect and the problems, risks and performativities associated with that production. Meta-effect might here be understood alongside Mark Hansen's observations of the primary technicity of all forms of embodiment, and in that sense, as an underscoring of the continuous (re)perception of the processual mode of the self-in-sensation, where subjectivity occurs as a characteristic of that which is 'always in movement, in variation'.[77] Meta-affect, then, as affectivity in process, might be understood as the experience of sensing affect in motion: the corporeal undoing of subjectivity such that we instead feel affect's corporeal coercions. As a form of *sensationship*, a form of spectatorship that hurts, it emerges here through a specific theatrical process that involves de-remediating affect from its remediating role in the media machine.

Hansen's and Bennett's emphases on those aesthetic forms which enable a re-perception of the function of everyday affects might be seen to offer a radical re-understanding of the way aesthetic forms enfold us as thinking, feeling, sensing beings, as well as pointing to the ways in which we are each corporeally aesthetic assemblages. For Laura Cull, theatre, in its most *affective* sense, *thinks*, in that it materialises the innate corporeality of sociality such that we might understand 'all experience, all material encounters, [as] already thoughtful, and... the activities that we are accustomed to referring to as "performance" and "philosophy" [as]... *attend[ing]* to the kinds of thinking that matter enacts on its own terms'.[78] Thinking theatre's thinking as a fundamental kind of feeling opens out exactly how sensuously complicitous we are in those other assemblages of visual and material culture that enfold our bodies in paradigms of representation that are better understood as a dynamics of exchange. In conceiving of aesthetics in materially affective terms, artworks become modes of encounter that work relationally upon, with, into and *as* spectator bodies. This turn to a sensory aesthetics produces a fundamentally different way of perceiving relations between object and subject, and thus a fundamentally different way of theorising the kinds of embodied and material thinking that objects can do. It is these theorisations that point out the implications of affect theory for empirical, or any other, epistemological platforms, including those that produce knowledges about cultures of memory.

Affect theory, while returning us to the body, also understands that the very productivity of that return exists in its contingencies. In bringing performance studies' methods of analysis to read memory cultures as they are engaged in practices of feeling, this book has been invested in a form of embodied speculation (and speculative embodiment) that devolves from how it stages in writing a series of relational encounters that are always already productive of those encounters as a kind of after-effect. Re-feeling here is both subject and object of the method of enquiry, and also not enough. The question of affect's spill, then, is less about Bennett's theorisation of a direct encounter with affect, and more about a sensing of the horizon of the memory affect as it enfolds spectators in the specific forms of sociality that it produces them through (and that it produces through them). Affect's spill, as the phenomenal revelation of the memory affect, is the perception of the self in the process of itself: the neither/and/or of affect on its way to (un)becoming the property and practice of cultures of memory. A work such as *BR.#04 Bruxelles* shows us the critical importance of noting how and when we feel ourselves feeling, and is one aesthetic means by which we might just vitally enliven our minds, senses and selves to other ways of doing, thinking, feeling and being.

Notes

Introduction: Feeling the Return of Memory

1. The Israeli Center for Digital Art, Artur Żmijewski, *80064* (2005), *The Israeli Center for Digital Art*, http://www.digitalartlab.org.il/ArchiveVideo.asp?id=16, page accessed 8 May 2013.
2. *Ibid.*, Poland.
3. Claire Bishop, *Artificial Hells: Participatory Art and the Politics of Spectatorship* (London: Verso, 2012), p. 353, 43*n*.
4. Elaine Scarry, *The Body in Pain: The Making and Unmaking of the World* (New York: Oxford University Press, 1985).
5. Sara Ahmed, *The Cultural Politics of Emotion* (New York: Routledge, 2004), p. 191.
6. Emily Apter argues that affect 'is what comes (quite logically) after the performative subject', in *Continental Drift: From National Characters to Virtual Subjects* (Chicago: University of Chicago Press, 1999), p. 20.
7. Ahmed, *The Cultural Politics of Emotion*, p. 6. Ahmed is working off the idea proposed by David Hume.
8. *Ibid.*, p. 31. Italics in original.
9. Scarry has written: 'there is no language for pain...it (more than any other phenomenon) resists verbal objectification', in *The Body in Pain*, p. 12.
10. Ahmed, *The Cultural Politics of Emotion*, p. 23.
11. Vicki Kirby, *Telling Flesh: The Substance of the Corporeal* (New York: Routledge, 1997), p. 65.
12. Judith Butler, 'Afterword: After Loss, What Then?' in David L. Eng and David Kazanjian (eds.) *Loss: The Politics of Mourning* (Berkeley: University of California Press, 2003), pp. 467–73, p. 467.
13. Jodi Rudoren, 'Proudly Bearing Elders' Scars, Their Skin Says "Never Forget"', *New York Times* New York Edition (1 October 2012), p. A1. I am grateful to Ilana Cohn for drawing my attention to this story.
14. Laurie Beth Clark, 'Always Already Again: Trauma Tourism and the Politics of Memory Culture', *Encounters* 1 (2010), pp. 65–74, p. 71.
15. Paul Connerton, *The Spirit of Mourning: History, Memory and the Body* (Cambridge: Cambridge University Press, 2011), p. 133. Italics in original.
16. *Ibid.*, p. ix. Italics in original.
17. *Ibid.*, p. 143. Italics in original.
18. Rebecca Schneider, *Performing Remains: Art and War in Times of Theatrical Reenactment* (Abingdon, Oxon: Routledge, 2011), pp. 1, 2.
19. Butler, 'Afterword: After Loss, What Then?', p. 467.
20. Judith Butler, *Precarious Life: The Powers of Mourning and Violence* (London: Verso, 2004), p. 23.

21. *Ibid.*, p. 22.
22. Lisa Saltzman, *Making Memory Matter: Strategies of Remembrance in Contemporary Art* (Chicago: University of Chicago Press, 2006).
23. Michael Hardt, 'What Affects Are Good For' in Patricia Ticineto Clough and Jean Halley (eds.) *The Affective Turn: Theorizing the Social* (Durham: Duke University Press, 2007), pp. ix–xiii, p. ix.
24. *Ibid.*, p. ix.
25. Patricia Ticineto Clough, 'Introduction' in Patricia Ticineto Clough and Jean Halley (eds.) *The Affective Turn: Theorizing the Social* (Durham: Duke University Press, 2007), pp. 1–33, p. 2.
26. *Ibid.*, p. 2.
27. Brian Massumi, *Parables for the Virtual: Movement, Affect, Sensation* (Durham: Duke University Press, 2002), p. 35. Italics in original.
28. Gregory J. Seigworth and Melissa Gregg, 'An Inventory of Shimmers' in Melissa Gregg and Gregory J. Seigworth (eds.) *The Affect Theory Reader* (Durham: Duke University Press, 2010), pp. 1–28, p. 13. Italics in original.
29. Massumi, *Parables for the Virtual*, p. 75.
30. *Ibid.*, p. 4.
31. Brian Massumi, 'The Future Birth of the Affective Fact: The Political Ontology of Threat' in Melissa Gregg and Gregory J. Seigworth (eds.) *The Affect Theory Reader* (Durham: Duke University Press, 2010), pp. 52–70, p. 65. Italics in original.
32. Eric Shouse, 'Feeling, Emotion, Affect', *M/C Journal* 8.6 (2005), para. 15. http://www.journal.media-culture.org.au/0512/03-shouse.php, page accessed 10 September 2013.
33. Anna Gibbs, 'Disaffected', *Continuum* 16.3 (2002), pp. 335–41, p. 337.
34. *Ibid.*, p. 338.
35. Teresa Brennan, *The Transmission of Affect* (Ithaca, New York: Cornell University Press, 2004), p. 1.
36. *Ibid.*, pp. 3, 7.
37. *Ibid.*, p. 25.
38. Clough, 'Introduction', p. 2.
39. See Jan Assmann, 'Collective Memory and Cultural Identity', trans. John Czaplicka *New German Critique* 65 (1995), pp. 125–33; Maurice Halbwachs *On Collective Memory*, trans. Lewis A. Coser (Chicago: University of Chicago Press, 1992 [1941]).
40. Richard Crownshaw, *The Afterlife of Holocaust Memory in Contemporary Literature and Culture* (Houndmills, Basingstoke: Palgrave, 2010), p. 3.
41. Wulf Kansteiner, 'Finding Meaning in Memory: A Methodological Critique of Collective Memory Studies', *History and Theory* 41 (May 2002), pp. 179–97, p. 182.
42. Susannah Radstone and Katharine Hodgkin (eds.) argue that the study of what they term '*regimes* of memory' opens out questions that foreground 'not memory's essence nor its ontology, but discursive *productions* of "memory"', in *Memory Cultures: Memory, Subjectivity and Recognition* (New Brunswick: Transaction, 2009), p. 1. Italics in original.

43. Andreas Huyssen, *Twilight Memories: Marking Time in a Culture of Amnesia* (New York: Routledge, 1995), pp. 5, 9.
44. Pierre Nora, 'Between Memory and History: *Les Lieux de Mémoire*' trans. Marc Roudebush *Representations* 26 (Spring 1989), pp. 7–25, p. 7.
45. Huyssen, *Twilight Memories*, p. 14.
46. Nora, 'Between Memory and History', p. 12.
47. That this affective space of transformation occurs in relation to both the 'intellectual and cultural traditions that frame all our representations of the past, [and] the memory makers who selectively adopt and manipulate these traditions' is paramount. Kansteiner, 'Finding Meaning in Memory', p. 180.
48. Schneider, *Performing Remains*, p. 33.
49. Crownshaw, *The Afterlife of Holocaust Memory in Contemporary Literature and Culture*, p. 2.
50. Marianne Hirsch, 'Surviving Images: Holocaust Photographs and the Work of Postmemory', *Yale Journal of Criticism* 14.1 (2001), pp. 5–37, p. 9.
51. Marianne Hirsch, *Family Frames: Photography, Narrative and Postmemory* (Cambridge, Massachusetts: Harvard University Press, 1997), p. 245.
52. Hirsch, 'Surviving Images', p. 9.
53. Kaja Silverman, *The Threshold of the Visible World* (New York: Routledge, 1996), p. 189. I would like to thank Caroline Wake for directing me to this text.
54. Chris Marker, *Sans Soleil* (Argos Films, France 1983).
55. Alison Landsberg, *Prosthetic Memory: The Transformation of American Remembrance in the Age of Mass Culture* (New York: Columbia University Press, 2004), p. 2.
56. *Ibid.*, p. 21.
57. Landsberg further adds: 'A practice of empathy is an essential part of taking on prosthetic memories, of finding ways to inhabit other people's memories *as* other people's memories and thereby respecting and recognizing difference,' in *Prosthetic Memory*, p. 24.
58. Ahmed, *The Cultural Politics of Emotion*, p. 30.
59. Lauren Berlant, 'The Subject of True Feeling: Pain, Privacy, and Politics' in Karyn Ball (ed.) *Traumatizing Theory: The Cultural Politics of Affect in and Beyond Psychoanalysis* (New York: Other Press, 2007), pp. 305–47, p. 309, p. 310.
60. *Ibid.*, p. 311.
61. Landsberg has been taken to task for her generality – what Crownshaw sees as the 'universalisation of affect or assumption of trauma' (*The Afterlife*, p. 24); and also for what James Berger describes as the literalisation of the term 'prosthetic'. Berger writes: 'Whatever it is we are experiencing or encountering in the events Landsberg describes, it is highly mediated...In a word, we encounter representations'. In 'Which Prosthetic? Mass Media, Narrative, Empathy, and Progressive Politics', *Rethinking History* 11.4 (2007), pp. 597–612, p. 604.
62. Dora Apel, *Memory Effects: The Holocaust and the Art of Secondary Witnessing* (New Brunswick: Rutgers, 2002), pp. 12, 31.

63. Nicholas Ridout, *Theatre & Ethics* (Houndmills, Basingstoke: Palgrave Macmillan, 2009), p. 67.
64. Berlant, 'The Subject', p. 310. My italics.
65. Amanda Wise, ' "It's Just an Attitude that You Feel": Inter-Ethnic Habitus Before the Cronulla Riots' in Greg Noble (ed.) *Lines in the Sand: The Cronulla Riots, Multiculturalism and National Belonging* (Sydney: Federation Press, 2009), pp. 127–45, p. 132.
66. Eve Kosofsky Sedgwick, *Touching Feeling: Affect, Pedagogy, Performativity* (Durham: Duke University Press, 2003), p. 19.
67. Ben Highmore, 'Bitter After Taste: Affect, Food, and Social Aesthetics' in Melissa Gregg and Gregory J. Seigworth (eds.) *The Affect Theory Reader* (Durham: Duke University Press, 2010), pp. 118–37, p. 120.
68. Highmore is drawing on the work of Alexander Baumgarten in 'Bitter After Taste', pp. 120, 121.
69. Jennifer Fisher discusses tactility, kinaesthetics and proprioception. What she terms relational aesthetics 'involves recovering a sensorially nuanced aesthetics in order to understand... the beholder's sensory production of knowledge', in 'Relational Sense: Towards a Haptic Aesthetics', *Parachute* 87 (July–September, 1997), pp. 4–11, pp. 4–5.
70. John Sutton *et al.* note that while there is a disjunction between the humanities' and scientific approaches to embodied cognition, their merging also raises problems, as it is 'difficult for humanities scholars to find the right scientific and psychological theories on which to draw and with which to seek articulations'. It is, however, equally deceptive to think that 'there are substantially unified visions of memory *within* either the sciences or the humanities'. See John Sutton, Celia B. Harris and Amanda J. Barnier, 'Memory and Cognition' in Susannah Radstone and Bill Schwarz (eds.) *Memory: Histories, Theories, Debates* (New York: Fordham University Press, 2010), pp. 209–26, p. 209. Italics in original.
71. Felicity Callard and Constantina Papoulias, 'Affect and Embodiment' in Susannah Radstone and Bill Schwarz (eds.) *Memory: Histories, Theories, Debates* (New York: Fordham University Press, 2010), pp. 246–62, p. 247.
72. As summarised by E.B. Titchener in 'Affective Memory', *The Philosophical Review* 4.1 (January 1895), pp. 65–76, p. 70.
73. *Ibid.*, p. 73.
74. Sutton *et al.*, 'Memory and Cognition', p. 211. See also John Sutton, Celia B. Harris, Paul G. Keil and Amanda J. Barnier, 'The Psychology of Memory, Extended Cognition, and Socially Distributed Remembering', *Phenomenology and the Cognitive Sciences* 9.4 (2010), pp. 521–60.
75. Edward S. Casey, *Remembering: A Phenomenological Study* 2nd edition (Bloomington: Indiana University Press, 2000 [1987]), p. 167. See also Casey, 'Habitual Body and Memory in Merleau-Ponty', *Man and World* 17 (1984), pp. 279–97.
76. *Ibid.*, p. 166.
77. *Ibid.*, pp. 167, 168.
78. Thomas J. Csordas, 'Somatic Modes of Attention', *Cultural Anthropology* 8.2 (1993), pp. 135–56, p. 138.

79. C. Nadia Seremetakis, 'The Memory of the Senses, Part 1: Marks of the Transitory' in C. Nadia Seremetakis (ed.) *The Senses Still: Perception and Memory as Material Culture in Modernity* (Chicago: University of Chicago Press, 1994), pp. 1–18, p. 9.
80. Antonio R. Damasio, *The Feeling of What Happens: Body and Emotion in the Making of Conciousness* (New York: Harcourt Brace and Co. 1999), p. 10.
81. Mary Helen Immordino-Yang and Antonio Damasio, 'We Feel, Therefore We Learn: The Relevance of Affective and Social Neuroscience to Education', *Mind, Brain and Education* 1.1 (2007), pp. 3–10, p. 8.
82. Damasio, *The Feeling of What Happens*, p. 279.
83. These perspectives are notably different to current disciplinary approaches which, as Susannah Radstone and Katherine Hodgkin observe, tend to differentiate between 'the possibility of genetic or biological inheritance, versus those for whom character and disposition, whether individual or collective, are shaped, in the main, by cultural and environmental forces', in 'Believing the Body' in Susannah Radstone and Katherine Hodgkin (eds.) *Regimes of Memory* (London: Routledge, 2003), pp. 23–6, p. 23.
84. Clough, 'Introduction', p. 14.

1 Sensing the Holocaust Affect: Memorials in Repeat, Revision and Return

1. Jane Korman notes on her website that after she removed the quotation marks from the original title 'Dancing Auschwitz', the clip 'went viral'. Jane Korman, *'Dancing Auschwitz Goes Viral', Jane Korman Art*. http://www.janekormanart.com/janekormanart.com/DA_Hype.html, page accessed 13 September 2013.
2. See reports in the following: Haaretz News Service, 'Jewish Artist Defends YouTube Video "Dancing Auschwitz"', *Haaretz News Service* (10 July 2010). http://www.haaretz.com/jewish-world/jewish-artist-defends-youtube-video-dancing-auschwitz-1.301096, page accessed 13 September 2013; Aimee Neistat, 'Dancing on the Ashes', *The Jerusalem Post.com* (10 August 2010). http://www.jpost.com/Magazine/Features/Article.aspx?id=190503, page accessed 13 September 2013; *The Daily Telegraph*, 'Outrage over Melbourne Artist Jane Korman's I Will Survive Dance at Polish Death Camp', *The Daily Telegraph* (14 July 2010). http://www.theaustralian.com.au/news/outrage-over-melbourne-artist-jane-kormans-i-will-survive-dance-at-polish-death-camp/story-e6frg6of-1225891392172, page accessed 13 September 2013. See also Korman's archive of media responses at *Jane Korman Art*. http://www.janekormanart.com/janekormanart.com/DA_Links.html, page accessed 13 September 2013.
3. The video installation *Dancing Auschwitz* comprises a large photographic image with three video pieces, respectively titled *I Will Survive: Dancing Auschwitz Parts 1, 2 and 3*. Korman explains that, for

maximum effect, the three video pieces are to be played together in a gallery space. In this chapter I solely refer to *I Will Survive: Dancing Auschwitz Part 1*, and for brevity's sake, refer to it as *Dancing Auschwitz*. The video work used in the installation is available at *Jane Korman Art*, http://www.janekormanart.com/janekormanart.com/16.Dancing_Auschwitz/16.Dancing_Auschwitz.html, page accessed 13 September 2013.
4. Vivian M. Patraka, *Spectacular Suffering: Theatre, Fascism and the Holocaust* (Bloomington: Indiana University Press, 1999). See also other recent studies which contemplate 'feeling' in relation to Holocaust history: Erika Doss, *Memorial Mania: Public Feeling in America* (Chicago: University of Chicago Press, 2010); and Gary Weissman, *Fantasies of Witnessing: Postwar Efforts to Experience the Holocaust* (Ithaca: Cornell University Press, 2004).
5. *Ibid.*, pp. 122, 121.
6. *Ibid.*, pp. 127, 110.
7. *Ibid.*, p. 6.
8. Vivian M. Patraka, 'Situating History and Difference: The Performance of the Term *Holocaust* in Public Discourse' in Jonathan Boyarin and Daniel Boyarin (eds.) *Jews and Other Differences: The New Jewish Cultural Studies* (Minneapolis: University of Minnesota Press, 1997), pp. 54–78, pp. 54–5. Patraka further explains that the Holocaust subsumes other acts of genocide to work 'as a kind of controlling or hegemonic discourse... that operates at the expense of the sufferings of other groups', p. 57.
9. Naomi Mandel, 'Rethinking "After Auschwitz": Against a Rhetoric of the Unspeakable in Holocaust Writing', *Boundary 2* 28.2 (2001), pp. 203–28, p. 204.
10. *Ibid.*, p. 218.
11. *Ibid.*, pp. 223–4.
12. Caroline Wake discusses the problematic application of the term 'witnessing' to a variety of spectatorial and experiential modes of relation to the event. See 'The Accident and the Account: Towards a Taxonomy of Spectatorial Witness in Theatre and Performance Studies', in Bryoni Trezise and Caroline Wake (eds.) *Visions and Revisions: Performance, Memory, Trauma* (Copenhagen: Museum Tusculanum Press, 2013), pp. 33–56.
13. Lauren Berlant, 'The Subject of True Feeling: Pain, Privacy, and Politics' in Karyn Ball (ed.) *Traumatizing Theory: The Cultural Politics of Affect in and Beyond Psychoanalysis* (New York: Other Press, 2007), pp. 305–47, p. 311.
14. Richard Crownshaw, *The Afterlife of Holocaust Memory in Contemporary Literature and Culture* (Houndmills, Basingstoke: Palgrave, 2010), p. vii.
15. Rebecca Schneider, *Performing Remains: Art and War in Times of Theatrical Reenactment* (Abingdon, Oxon: Routledge, 2011), p. 6.
16. *Ibid.*, p. 14. Italics in original.
17. I previously wrote about this tension with Caroline Wake in Bryoni Trezise and Caroline Wake, 'Introduction to After Effects: Performing the Ends of Memory', *Performance Paradigm* 5.1 (May

2009). http://www.performanceparadigm.net/wp-content/uploads/2009/07/wake-and-trezise-intro-final.pdf, page accessed 13 September 2013.
18. Postmemory attempts 'to define both a specifically inter- and trans-generational act of transfer and the resonant aftereffects of trauma', in Marianne Hirsch, 'The Generation of Postmemory', *Poetics Today* 29.1 (2008), pp. 103–28, p. 106.
19. Barbara Kirshenblatt-Gimblett, *Destination Culture: Tourism, Museums, and Heritage* (Berkeley: University of California Press, 1998), p. 20.
20. *Ibid.*, p. 18, p. 20.
21. Schneider, *Performing Remains*, p. 102; Peggy Phelan, *Unmarked: The Politics of Performance* (London: Routledge, 1993), p. 19. Italics in original.
22. *Ibid.*, p. 13.
23. 'Performance cannot...participate in the circulation of representations *of* representations', Phelan writes, 'once it does so, it becomes something other than performance', in *Unmarked*, p. 146. Italics in original.
24. Peggy Phelan, *Mourning Sex: Performing Public Memories* (New York: Routledge, 1997), p. 4.
25. Paul Stoller, *Embodying Colonial Memories: Spirit Possession, Power, and the Hauka in West Africa* (New York: Routledge, 1995), p. 21.
26. Joseph Roach, *Cities of the Dead: Circum-Atlantic Performance* (New York: Columbia University Press, 1996), p. 5.
27. Diana Taylor, *The Archive and the Repertoire: Performing Cultural Memory in the Americas* (Durham: Duke University Press, 2003), p. 20.
28. *Ibid.*, p. 86.
29. Schneider, *Performing Remains*, pp. 104, 102.
30. See Susan Leigh Foster, *Choreographing Empathy: Kinesthesia in Performance*, (Abingdon, Oxon: Routledge, 2011).
31. See Geraldine Harris, 'The Ethics and Politics of Witnessing Whoopi' in Bryoni Trezise and Caroline Wake (eds.) *Visions and Revisions: Performance, Memory, Trauma* (Copenhagen: Museum Tusculanum Press, 2013), pp. 115–30.
32. And further: 'When I feel angry, I feel the passage of anger through me. What I feel with and what I feel are distinct', in Teresa Brennan, *The Transmission of Affect* (Ithaca, New York: Cornell University Press, 2004), p. 5.
33. *Ibid.*, p. 6.
34. *Ibid.*, p. 9.
35. *Ibid.*, p. 10.
36. Mark Dapin, 'Lest We Remember', *The Sydney Morning Herald Good Weekend* (8 July 2006), pp. 33–4, p. 34. He continues: 'I am not sure it should be a whistlestop on a world tour, sandwiched between the Hofbräuhaus and the Frauenkirche on the weekend after Oktoberfest', p. 34.
37. *Ibid.*, p. 34.
38. Dachau Concentration Camp Memorial Site, 'Virtual Tour', *Dachau Concentration Camp Memorial Site*. http://www.kz-gedenkstaette-dachau.de/virtual_tour.html, page accessed 13 September 2013.

39. Dora Apel, *Memory Effects: The Holocaust and the Art of Secondary Witnessing* (New Brunswick: Rutgers, 2002), p. 31.
40. Weissman, *Fantasies of Witnessing*, p. 20.
41. *Ibid.*, p. 21. Weissman is quoting Ellen S. Fine.
42. *Ibid.*, p. 26. Italics in original.
43. *Ibid.*, p. 21.
44. Omer Bartov, 'Defining Enemies, Making Victims: Germans, Jews, and the Holocaust', *The American Historical Review* 103.3 (1998), pp. 771–816, pp. 771–2.
45. Dominick LaCapra, *History and Memory After Auschwitz* (Ithaca: Cornell University Press, 1998), p. 34.
46. Mandel, 'Rethinking "After Auschwitz" ', pp. 223–4.
47. Weissman, *Fantasies of Witnessing*, p. 211.
48. Abigail Solomon-Godeau, 'Mourning or Melancholia: Christian Boltanski's "Missing House" ', *Oxford Art Journal* 21.2 (1998), pp. 3–20, p. 3.
49. As quoted in Marianne Hirsch, *Family Frames: Photography, Narrative and Postmemory* (Cambridge, Massachusetts: Harvard University Press, 1997), p. 259.
50. Steven L. Sampson, 'From Reconciliation to Coexistence', *Public Culture* 15.1 (2003), pp. 181–6, p. 183.
51. Susan Best, 'Rethinking Visual Pleasure: Aesthetics and Affect', *Theory Psychology* 17.4 (2007), pp. 505–14, p. 511.
52. Brian Massumi, *Parables for the Virtual: Movement, Affect, Sensation* (Durham: Duke University Press, 2002), p. 35.
53. *Ibid.*, p. 28.
54. *Ibid.*, p. 32.
55. *Ibid.*, p. 31. Italics in original.
56. Clare Hemmings, 'Invoking Affect', *Cultural Studies* 19.5 (2005), pp. 548–67, p. 554.
57. Brennan, *The Transmission of Affect*, pp. 25, 3.
58. Hemmings, 'Invoking Affect', p. 551.
59. The study they cite involves a blind taste experiment in which participants drank Pepsi masked as Coke; there was a preference for the drink 'tagged' as Coke despite the control group preferring each drink equally. Jeff Pruchnic and Kim Lacey, 'The Future of Forgetting: Rhetoric, Memory, Affect', *Rhetoric Society Quarterly* 41.5 (2011), pp. 472–94, p. 488.
60. Further, as they observe via neuroscientist Antonio Damasio, 'affective responses are stored in dormant and implicit "dispositional memories" that record these responses within our nervous systems'. Pruchnic and Lacey, 'The Future of Forgetting', p. 486.
61. Massumi, *Parables for the Virtual*, p. 35.
62. Solomon-Godeau, 'Mourning or Melancholia', p. 3.
63. As quoted in Solomon-Godeau, 'Mourning or Melancholia', p. 7.
64. The Jewish Museum Berlin, 'The Installations', *The Jewish Museum Berlin*. http://www.jmberlin.de/main/EN/01-Exhibitions/04-installations.php, page accessed 14 September 2013.
65. *The Scream of Nature* was the original title given to the work.

66. Fredric Jameson, *Postmodernism, or, the Cultural Logic of Late Capitalism* (Durham: Duke University Press, 2001 [1991]), p. 10.
67. Ibid., p. 19.
68. Tomkins cited in Anna Gibbs, 'Disaffected', *Continuum* 16.3 (2002), pp. 335–41, p. 338.
69. Ibid., p. 338.
70. Emil Hrvatin, 'The Scream', *Performance Research* 2.1 (1997), pp. 82–91, p. 87.
71. Ibid., p. 88. My italics.
72. See: David A. Ellison, 'The Spoiler's Art: Embarrassed Space as Memorialisation', *The South Atlantic Quarterly* 110.1 (2011), pp. 89–100; Quentin Stevens, 'Nothing More than Feelings: Abstract Memorials', *Architectural Theory Review* 14.2 (2009), pp. 156–72; Irit Dekel, 'Ways of Looking: Observation and Transformation at the Holocaust Memorial, Berlin', *Memory Studies* 2.1 (2009), pp. 71–86, p. 74.
73. Joachim Schlör, *Memorial to the Murdered Jews in Europe*, trans. Paul Aston (Munich: Prestel, 2005), p. 38.
74. As cited in Jason Hollander, 'Peter Eisenman, Architecture '60, Designs New Holocaust Memorial in Berlin', *Columbia News* (10 July 2003). http://www.columbia.edu/cu/news/03/07/peterEisenman.html, page accessed 13 September 2013.
75. Edward, 'Comment' in John Hill, *A Daily Dose of Architecture* (11 May 2005). http://www.archidose.blogspot.com/2005/05/stelae.html, page accessed 13 September 2013.
76. Eisenman quoted in BBC World News, 'Berlin Opens Holocaust Memorial', *BBC World News* (10 May 2005). http://news.bbc.co.uk/2/hi/4531669.stm, page accessed 13 September 2013.
77. Schlör, *Memorial to the Murdered Jews in Europe*, p. 5.
78. Ibid., p. 38.
79. Ibid., p. 33.
80. James E. Young, 'The Counter-Monument: Memory Against Itself in Germany Today', *Critical Inquiry* 18.2 (1992), pp. 267–96. See also Mark Godfrey, *Abstraction and the Holocaust* (New Haven: Yale University Press, 2007).
81. Ibid., p. 296.
82. Lisa Saltzman, *Making Memory Matter: Strategies of Remembrance in Contemporary Art* (Chicago: University of Chicago Press, 2006), p. 13.
83. Joan Simon, *Ann Hamilton: An Inventory of Objects* (New York: Gregory R. Miller and Co., 2006), p. 157.
84. Saltzman, *Making Memory Matter*, p. 10.
85. Jane Korman, 'Dancing Auschwitz', *Jane Korman Art*. http://www.janekormanart.com/janekormanart.com/16.Dancing_Auschwitz/Pages/I_Will_Survive_Video.html, page accessed 13 September 2013.
86. Adolek Kohn, as quoted in Jane Korman, *I Will Survive: Dancing Auschwitz Part 1*.
87. Sara Ahmed, 'Happy Objects' in Melissa Gregg and Gregory J. Seigworth (eds.) *The Affect Theory Reader* (Durham: Duke University Press, 2010), pp. 29–51, p. 36.

2 Becoming Other-wise: Remembering Intercorporeal Indigeneity *Down Under*

1. See Ghassan Hage, *Against Paranoid Nationalism: Searching for Hope in a Shrinking Society* (Sydney: Pluto Press, 2003); and Suvendrini Perera and Joseph Pugliese, ' "Racial Suicide": The Re-licensing of Racism in Australia', *Race Class* 39.1 (1997), pp. 1–19.
2. Chris Healy, *Forgetting Aborigines* (Sydney: UNSW Press, 2008), p. 5. Healy is building on the words of Marcia Langton.
3. Stephen Muecke, 'Lonely Representations: Aboriginality and Cultural Studies', *Journal of Australian Studies* 16.35 (1992), pp. 32–44, p. 43. Italics in original.
4. Sara Ahmed, 'The Non-Performativity of Anti-Racism', *Borderlands e-journal* 5.3 (2005). http://www.borderlands.net.au/vol5no3_2006/ahmed_nonperform.htm, page accessed 30 August 2013.
5. Elizabeth A. Povinelli, *The Cunning of Recognition: Indigenous Alterities and the Making of Australian Multiculturalism* (Durham: Duke University Press, 2002), pp. 5–6. Ghassan Hage considers these effects through the now well-established multicultural food fair that emerged in 1980s multicultural policies. See *White Nation: Fantasies of White Supremacy in a Multicultural Society* (New York: Routledge, 2000), p. 117.
6. See Sneja Gunew, *Haunted Nations: The Colonial Dimensions of Multiculturalisms* (Abingdon, Oxon, Routledge: 2004).
7. Povinelli, *The Cunning of Recognition*, p. 6.
8. Kelly Jean Butler, *Witnessing Australian Stories: History, Testimony, and Memory in Contemporary Culture* (New Jersey: Transaction Publishers, 2013), p. 25.
9. *Ibid.*, p. 22.
10. *Ibid.*, p. 25.
11. David Rowe and Deborah Stevenson, 'Sydney 2000: Sociality and Spatiality in Global Media Events' in Alan Tomlinson and Christopher Young (eds.) *National Identity and Global Sports Events: Culture, Politics, and Spectacle in the Olympics and the Football World Cup* (Albany: State University of New York Press, 2006), pp. 197–214, p. 197.
12. These words have been transcribed by the author from the televised live commentaries given respectively by Bruce McAveney and Ernie Dingo during the Sydney 2000 Olympics Opening Ceremony, broadcast by Channel 7, 15 September 2000.
13. Michael Cohen, Paul Dwyer and Laura Ginters, 'Performing "Sorry Business": Reconciliation and Redressive Action' in Graham St John (ed.) *Victor Turner and Contemporary Cultural Performance* (Oxford: Berghahn Books, 2008), pp. 76–93, p. 85.
14. *Ibid.*, p. 85.
15. *Ibid.*, p. 86.
16. Rowe and Stevenson, 'Sydney 2000', p. 202.
17. Catriona Elder, Angela Pratt and Cath Ellis, 'Running Race: Reconciliation, Nationalism and the Sydney 2000 Olympic Games', *International*

Review for the Sociology of Sport 41.2 (2006), pp.181–200, p. 182. Italics in original.
18. *Ibid.*, p. 181.
19. Diana Taylor, *The Archive and the Repertoire: Performing Cultural Memory in the Americas* (Durham: Duke University Press, 2003), p. 28.
20. *Ibid.*, p. 29.
21. Ahmed, 'The Non-Performativity of Anti-Racism'. My italics.
22. Taylor, *The Archive and the Repertoire*, p. 13.
23. Povinelli, *The Cunning of Recognition*, p. 6.
24. *Ibid.*, p. 48.
25. *Ibid.*, p. 50.
26. Povinelli argues that there is a perception that 'traditional Indigenous culture has a different relationship to national time and space', *The Cunning of Recognition*, p. 48. Stephen Muecke also raises this point in *Ancient and Modern: Time, Culture and Indigenous Philosophy* (Sydney: UNSW Press, 2004), p. 15.
27. *Ibid.*, p. 34.
28. *Ibid.*, p. 50.
29. Roslyn Poignant, *Professional Savages: Captive Lives and Western Spectacle* (New Haven: Yale University Press, 2004).
30. *Ibid.*, p. 7.
31. *Ibid.*, pp. 4, 6.
32. *Ibid.*, p. 8.
33. *Ibid.*, p. 4.
34. Jane R. Goodall, *Performance and Evolution in the Age of Darwin: Out of the Natural Order* (London: Routledge, 2002), p. 89.
35. Poignant, *Professional Savages*, p. 7.
36. Michael Parsons, 'The Tourist Corroboree in South Australia to 1911', *Aboriginal History* 21 (1997), pp. 46–69, p. 46, p. 47–8.
37. *Ibid.*, p. 47.
38. Michael Parsons, ' "Ah That I Could Convey a Proper Idea of this Interesting Wild Play of the Natives": Corroborees and the Rise of Indigenous Australian Cultural Tourism', *Australian Aboriginal Studies*, 2 (2002), pp. 14–26, p. 17.
39. Muecke, 'Lonely Representations', p. 40. Italics in original.
40. Former Indigenous Tourism Australia Executive Chairman Aden Ridgeway quoted in Andrew Bain, 'Destination Dreaming', *Australian Geographic* 88 (October–December 2007), pp. 72–85, p. 75.
41. *Ibid.*, p. 77.
42. *Ibid.*, p. 77.
43. Bain, 'Destination Dreaming', p. 77. See also Chris Ryan and Jeremy Huyton, 'Tourists and Aboriginal People', *Annals of Tourism Research* 29.3, pp. 631–47, p. 636.
44. Ryan and Huyton, 'Tourists and Aboriginal People', p. 633.
45. Rosita Henry, 'Dancing Into Being: The Tjapukai Aboriginal Cultural Park and the Laura Dance Festival', *Australian Journal of Anthropology* 11.3 (2000), pp. 322–32, p. 329. This might be considered in comparison to

Chris Healy's discussion of walking the Lurujarri Heritage Trail north of Broome in *Forgetting Aborigines*, pp. 181–202.
46. As discussed in Pam Dyer, Lucinda Aberdeen and Sigrid Schuler, 'Tourism Impacts on an Australian Indigenous Community: A Djapugay Case Study', *Tourism Management* 24 (2004), pp. 83–95, p. 83.
47. Tjapukai Aboriginal Cultural Park, 'Tjapukai 25 Years On'. http://www.tjapukai.com.au/, page accessed 15 September 2013.
48. *Ibid.*
49. Dyer *et al.*, 'Tourism Impacts', p. 88. Dyer *et al.* conducted interviews, which grew out of extensive consultation with the Djapugay people, over a six-month period.
50. *Ibid.*, p. 90.
51. Tjapukai Aboriginal Cultural Park, 'Night Tour'. http://www.tjapukai.com.au/, page accessed 15 September 2013.
52. Nicholas Ridout, *Stage Fright, Animals, and Other Theatrical Problems*, (Cambridge: Cambridge University Press, 2006), p. 3, quoting Jonas Barish; and p. 33.
53. *Ibid.*, pp. 33, 34.
54. Recalling Povinelli, *The Cunning of Recognition*, p. 50.
55. Parsons, 'Corroborees and the Rise of Indigenous Australian Cultural Tourism', p. 18.
56. Parsons, 'The Tourist Corroboree in South Australia to 1911', p. 52.
57. Moorhouse wrote: 'I have told them repeatedly not to corroobory on the Sabbath, but crowds of Europeans visit them on this day, and offer them money, and the Natives find it more profitable to listen to them than to me. Last Sunday I believe they made nearly two pounds by their performances', as quoted in Parsons, 'Tourist Corroboree', p. 49.
58. Parsons, 'The Tourist Corroboree in South Australia to 1911', p. 53.
59. Maryrose Casey, 'Carnivalising Sovereignty: Containing the Indigenous Protest within the "White" Australian Nation', *About Performance* 7 (2007), pp. 69–84, p. 75.
60. Marcia Langton, 'Earth, Wind, Fire, and Water: The Social and Spiritual Construction of Water in Aboriginal Societies' in Bruno David, Bryce Barker and Ian J. McNiven (eds.) *The Social Archaeology of Australian Indigenous Societies* (Canberra: Aboriginal Studies Press, 2006), pp. 139–59, p. 139, p. 159. Langton elsewhere explains that ritual fire can 'warm' the bodies of visitors to country, as in the Lamalama's account of fire as a 'reciprocal ritual action that is undertaken by guests...in someone else's country', where the smoke clears away 'any spiritually ambient presence that the person has carried from his own country or among his personal collection of things', in 'The Edge of the Sacred, the Edge of Death: Sensual Inscriptions' in Bruno David and Meredith Wilson (eds.) *Inscribed Landscapes: Marking and Making Place* (Honolulu, HI: University of Hawai'i Press, 2002), pp. 253–69, pp. 263–4.
61. Simon Werrett, *Fireworks: Pyrotechnic Arts and Sciences in European History* (Chicago: University of Chicago Press, 2010), p. 3. My italics.

62. Patrick Wolfe, 'On Being Woken Up: The Dreamtime in Anthropology and in Australian Settler Culture', *Comparative Studies in Society and History* 33.2 (1991), pp. 197–224, p. 198.
63. *Ibid.*, pp. 204, 205.
64. *Ibid.*, p. 205.
65. *Ibid.*, p. 206.
66. *Ibid.*, p. 210.
67. Gillian Cowlishaw, 'Mythologising Culture Part 1: Desiring Aboriginality in the Suburbs', *Australian Journal of Anthropology* 21 (2010), pp. 208–27, p. 212.
68. *Ibid.*, p. 217.

3 Feeling Remediated: The Emotional Afterlife of Psychic Trauma TV

1. John Edward, *Crossing Over: The Stories Behind the Stories* (New York: Princess Books, 2001), p. xv.
2. Edward is quoting marketing guru Rick Korn, *Crossing Over*, p. 159.
3. Kevin Christopher, ' "I Speak to Dead People": Medium John Edward Hosts SciFi Cable Show', *Skeptical Inquirer* 24.5 (2000), p. 9.
4. Avery F. Gordon, *Ghostly Matters: Haunting and the Sociological Imagination* (Minneapolis: University of Minnesota Press, 1997), p. 201.
5. *Ibid.*, p. 17.
6. Excerpt quoted in Christopher, ' "I Speak to Dead People" ', p. 9.
7. John Hockenberry quoted in Edward, *Crossing Over*, p. 243.
8. Leon Jaroff, 'Talking to the Dead', *Time* 157.9 (5 March 2001), p. 52.
9. Edward, *Crossing Over*, pp. 5, 6.
10. *Ibid.*, p. 7.
11. Tom Gliatto and Natasha Stoynoff, 'Medium Rare', *People Weekly* 57.17 (6 May 2002), pp. 85–6, p. 85.
12. Jaroff, 'Talking to the Dead', p. 52.
13. Edward, *Crossing Over*, p. 238.
14. Gordon, *Ghostly Matters*, p. 8.
15. *Ibid.*, p. 8.
16. John Potts, 'The Idea of the Ghost' in John Potts and Edward Scheer (eds.) *Technologies of Magic: A Cultural Study of Ghosts, Machines and the Uncanny* (Sydney: Power Publications, 2006), pp. 78–91.
17. Gordon, *Ghostly Matters*, pp. 13–14.
18. Jeffrey Sconce, *Haunted Media: Electronic Presence from Telegraphy to Television* (Durham: Duke University Press, 2000), pp. 49–50. See also Erik Davis, *TechGnosis: Myth, Magic & Mysticism in the Age of Information* (New York: Three Rivers Press, 1998).
19. Edward, *Crossing Over*, p. xx.
20. Arena TV Australia, *Crossing Over* (November 2005). Author's transcription.
21. Alison Landsberg, 'Memory, Empathy, and the Politics of Identification', *International Journal of Politics, Culture and Society* 22 (2009), pp. 221–9, p. 222.

22. Gordon, *Ghostly Matters*, p. 13.
23. Sara Ahmed, 'Happy Objects' in Melissa Gregg and Gregory J. Seigworth (eds.) *The Affect Theory Reader* (Durham: Duke University Press, 2010), pp. 29–51, p. 40.
24. Brian Massumi, 'The Future Birth of the Affective Fact: The Political Ontology of Threat' in Melissa Gregg and Gregory J. Seigworth (eds.) *The Affect Theory Reader* (Durham: Duke University Press, 2010), pp. 52–70, p. 63, p. 65.
25. *Ibid.*, p. 63.
26. Richard Grusin, 'Remediation and Premediation', *Criticism* 46.1 (2004), pp. 17–40, p. 23.
27. *Ibid.*, p. 28.
28. Arena TV Australia, *Crossing Over* (November 2005). Author's transcription.
29. Jon Dovey, *Freakshow: First Person Media and Factual Television* (London: Pluto Press, 2000), p. 10.
30. Nigel Thrift, *Non-Representational Theory: Space, Politics, Affect* (Hoboken: Routledge, 2007), p. 184.
31. Dovey, *Freakshow*, pp. 10, 12.
32. *Ibid.*, pp. 21–2.
33. Arena TV Australia, *Crossing Over* (November 2005). Author's transcription.
34. Laura Grindstaff, *The Money Shot: Trash, Class and the Making of TV Talk Shows* (Chicago: University of Chicago Press, 2002).
35. Gliatto and Stoynoff, 'Medium Rare', p. 85.
36. Josh Wolk, 'Tomb Reader', *Entertainment Weekly* 614 (14 September 2001), pp. 57–9, p. 57.
37. As quoted in Edward, *Crossing Over*, p. 222.
38. Dovey, *Freakshow*, p. 97.
39. *Ibid.*, p. 26.
40. Arena TV Australia, *Crossing Over* (November 2005). Author's transcription.
41. Shermer quoted in Wolk, 'Tomb Reader', p. 58.
42. Edward, *Crossing Over*, p. 279.
43. Mary Marshall Clark, 'The September 11, 2001, Oral History Narrative and Memory Project: A First Report', *The Journal of American History* 89.2 (September 2002), pp. 569–79, p. 569.
44. Slavoj Zizek, *Welcome to the Desert of the Real: Five Essays on September 11 and Related Dates* (London: Verso, 2002), p. 13.

4 Affecting Indifference: Traumatic A-materiality in Second Life

1. Anne Frank Museum, 'The Secret Annex Online'. http://www.annefrank.org/en/Subsites/Home/; Auschwitz-Birkenau Memorial and Museum, 'Virtual Tour of Auschwitz Sites'. http://en.auschwitz.org/z/index.php?option=com_content&task=view&id=6&Itemid=8; University of Southern California Shoah Foundation, 'IWitness'. http://iwitness.usc.edu/SFI/.

See also, for an interesting inversion of the visibility of traumatic memory in these sites, Jewish Museum Berlin, 'What We Won't Show You'. http://www.jmberlin.de/osk/wwnz/filme/en/film-7/film.php, all pages accessed 17 September 2013.
2. See, respectively, Kate Douglas, 'Cyber-Commemoration: Life Writing, Trauma and Memorialisation', *Life Writing Symposium* 13–15 June 2006, Flinders University; Paul Longley Arthur, 'Exhibiting History: The Digital Future', *reCollections: Journal of the National Museum of Australia* 3.1 (2008), pp. 33–50; Giorgia Doná, 'Collective Suffering and Cyber-Memorialisation in Post-Genocide Rwanda' in Mick Broderick and Antonio Traverso (eds.) *Trauma, Media, Art* (Newcastle Upon Tyne: Cambridge Scholars Publishing: 2010), pp. 16–35.
3. See, in particular, Joanne Garde-Hansen, Andrew Hoskins and Anna Reading (eds.) *Save As...Digital Memories* (Houndmills, Basingstoke: Palgrave Macmillan, 2009); Andrew Hoskins, 'Anachronisms of Media, Anachronisms of Memory: From Collective Memory to a New Memory Ecology' in Motti Neiger, Oren Meyers and Eyal Zandberg (eds.) *On Media Memory: Collective Memory in a New Media Age* (New York: Palgrave Macmillan, 2011), pp. 278–88; Jose Van Dijck, *Mediated Memories in the Digital Age* (Stanford: Stanford University Press, 2007).
4. Andreas Huyssen, *Twilight Memories: Marking Time in a Culture of Amnesia* (New York: Routledge, 1995).
5. Vivian M. Patraka, *Spectacular Suffering: Theatre, Fascism and the Holocaust* (Bloomington: Indiana University Press, 1999).
6. United States Holocaust Memorial Museum, 'Witnessing History: Kristallnacht – The November 1938 Pogroms'. http://snurl.com/7r1i6, page accessed 17 September 2013.
7. Wai Chee Dimock, 'Introduction: Genres as Fields of Knowledge', *PMLA* 122.5 (2007), pp. 1377–88, p. 1380. Italics in original.
8. Gillian Whitlock, 'Remediating Gorilla Girl: Rape Warfare and the Limits of Humanitarian Storytelling', *Biography* 33.3 (2010), pp. 471–97, p. 472.
9. Julian Dibbell, 'A Rape in Cyberspace', *The Village Voice*, 21 December 1993, pp. 36–42, p. 37.
10. *Ibid.*, p. 37.
11. *Ibid.*
12. *Ibid.*, p. 38.
13. *Ibid.*, p. 37.
14. *Ibid.*
15. Theresa, M. Senft, 'Introduction: Performing the Digital Body – A Ghost Story', *Women & Performance: A Journal of Feminist Theory* 9.1 (1996), pp. 9–33, p. 17.
16. *Ibid.*, pp. 13, 14.
17. See Alex Burns and Ben Eltham, 'Twitter Free Iran: An Evaluation of Twitter's Role in Public Diplomacy and Information Operations in Iran's 2009 Election Crisis', *Communications Policy & Research Forum*, 19–20 November 2009, University of Technology, Sydney. http://eprints.vu.edu.au/15230/, page accessed 13 September 2013; and Mette Mortensen,

'When Citizen Photojournalism Sets the News Agenda: Neda Agha Soltan as a Web 2.0 Icon of Post-Election Unrest in Iran', *Global Media and Communication* 7.1 (2011), pp. 4–16.
18. Anna Reading, 'Memory and Digital Media: Six Dynamics of the Globital Memory Field' in Motti Neiger, Oren Meyers and Eyal Zandberg (eds.) *On Media Memory: Collective Memory in a New Media Age* (New York: Palgrave Macmillan, 2011), pp. 241–52, pp. 241–2.
19. *Ibid.*, p. 242.
20. Cathy Caruth, *Unclaimed Experience: Trauma, Narrative, and History* (Baltimore: The Johns Hopkins University Press, 1996), p. 4.
21. *Ibid.*, p. 17.
22. *Ibid.*, p. 18.
23. Avery F. Gordon, *Ghostly Matters: Haunting and the Sociological Imagination* (Minneapolis: University of Minnesota Press, 1997), p. 22. My italics.
24. Jay David Bolter and Richard Grusin, *Remediation: Understanding New Media* (Massachusetts: MIT Press, 2000 [1999]), p. 56.
25. Joanne Garde-Hansen, Andrew Hoskins and Anna Reading, 'Introduction' in Joanne Garde-Hansen, Andrew Hoskins and Anna Reading (eds.) *Save As...Digital Memories* (Houndmills, Basingstoke: Palgrave Macmillan, 2009), pp. 1–21, p. 14.
26. Astrid Erll and Ann Rigney, 'Introduction: Cultural Memory and its Dynamics' in Astrid Erll and Ann Rigney (eds.) *Mediation, Remediation and the Dynamics of Cultural Memory* (Berlin: Walter de Gruyter, 2009), pp. 1–14, p. 5. Italics in original.
27. Bolter and Grusin, *Remediation*, p. 21.
28. *Ibid.*, p. 56.
29. Marc Redfield, 'Virtual Trauma: The Idiom of 9/11', *Diacritics* 37.1 (2007), pp. 55–80, p. 68.
30. *Ibid.*, p. 59.
31. *Ibid.*, p. 61.
32. *Ibid.*, p. 75.
33. Ann Cvetkovich and Ann Pellegrini, 'Introduction', *S&F Online* 2.1 (2003). http://sfonline.barnard.edu/ps/intro.htm, page accessed 17 September 2013.
34. Sara Ahmed, *Strange Encounters: Embodied Others in Post-Coloniality* (London: Routledge, 2000), p. 8. Italics in original.
35. E. Ann Kaplan, *Trauma Culture: The Politics of Terror and Loss in Media and Literature* (New Brunswick: Rutgers, 2005), p. 87.
36. Mark Seltzer, *Serial Killers: Death and Life in America's Wound Culture* (New York: Routledge, 1998), p. 254.
37. Caroline Wake uses this term in specific relation to theories of witness: 'when witnessing a performance the spectator experiences a sort of "after-affect" rather than simply experiencing affect during the performance or the after-effects of that affect. The affect itself does not arrive during the performance but afterwards'. See 'The Accident and the Account: Towards a Taxonomy of Spectatorial Witness in Theatre and Performance Studies'

in Bryoni Trezise and Caroline Wake (eds.) *Visions and Revisions: Performance, Memory, Trauma* (Copenhagen: Museum Tusculanum Press, 2013), pp. 33–56, p. 38.
38. Cannon Schmitt, 'Introduction: Materia Media', *Criticism* 46.1 (2004), pp. 11–15, p. 11.
39. Richard Urban, Paul Marty and Michael Twidale, 'A Second Life for Your Museum: 3D Multi-User Virtual Environments and Museums' in *Museums and the Web 2007: The International Conference for Culture and Heritage On-line*, 11–14 April 2007, San Francisco, California. http://www.museumsandtheweb.com/mw2007//papers/urban/urban.html, page accessed 18 September 2013.
40. *Ibid.*
41. Naomi Mandel, 'Rethinking "After Auschwitz": Against a Rhetoric of the Unspeakable in Holocaust Writing', *Boundary 2* 28.2 (2001), pp. 203–28.
42. Patricia Ticineto Clough, 'Introduction' in Patricia Ticineto Clough and Jean Halley (eds.) *The Affective Turn: Theorizing the Social* (Durham: Duke University Press, 2007), pp. 1–33, p. 2.
43. Brian Massumi, *Parables for the Virtual: Movement, Affect, Sensation* (Durham: Duke University Press, 2002), p. 75.
44. *Ibid.*, p. 30. Italics in original.
45. Clough, 'Introduction', p. 25.
46. *Ibid.*, pp. 25–6.
47. Mark B.N. Hansen, *Bodies in Code: Interfaces with Digital Media* (New York: Routledge, 2006), p. 5. Italics in original.
48. Antonio R. Damasio, *The Feeling of What Happens: Body and Emotion in the Making of Conciousness* (New York: Harcourt Brace and Co., 1999), p. 279.
49. Hansen, *Bodies in Code*, p. 8.
50. Emily Apter, *Continental Drift: From National Characters to Virtual Subjects* (Chicago: University of Chicago Press, 1999), p. 20.
51. *Ibid.*, p. 19.
52. Hansen, *Bodies in Code*, p. 122.
53. *Ibid.*, p. 147.
54. *Ibid.*, p. 276, nxiii.
55. David Klevan, Education Manager for Technology and Distance Learning Initiatives at the Division of Outreach Technology at the United States Holocaust Memorial Museum, who worked on the project, quoted in Joey Seiler, 'Holocaust Museum Launching Kristallnacht Second Life Exhibit with Involve', *Engage Digital*, 5 December 2008. www.engagedigital.com/2008/12/05/holocaust-museum-launching-kristallnacht-second-life-exhibit-with-involve/, page accessed 18 September 2013.
56. Barbara Kirshenblatt-Gimblett, *Destination Culture: Tourism, Museums, and Heritage* (Berkeley: University of California Press, 1998), p. 20.
57. *Ibid.*, p. 21.
58. Jeff Pruchnic and Kim Lacey, 'The Future of Forgetting: Rhetoric, Memory, Affect', *Rhetoric Society Quarterly* 41.5, pp. 472–94, p. 488.
59. Cited in Pruchnic and Lacey, 'The Future of Forgetting', p. 486.

60. Maria Tumarkin, *Traumascapes: The Power and Fate of Places Transformed by Tragedy* (Carlton: Melbourne University Press, 2005), p. 12.
61. Lisa Saltzman, *Making Memory Matter: Strategies of Remembrance in Contemporary Art* (Chicago: University of Chicago Press, 2006), p. 12.
62. *Ibid.*, pp. 3, 13.
63. Marianne Hirsch, 'Surviving Images: Holocaust Photographs and the Work of Postmemory', *Yale Journal of Criticism* 14.1 (2001), pp. 5–37.
64. Kirshenblatt-Gimblett, *Destination Culture*, p. 17.
65. This idea might be connected to Caroline Wake's characterisation of 'hypermediate witnessing', in which 'spectators [of video testimonies] are spatiotemporally distant and experience themselves as such because the medium does not recede but rather remains in view'. See 'Regarding the Recording: The Viewer of Video Testimony, the Complexity of Copresence and the Possibility of Tertiary Witnessing', *History and Memory* 25.1 (2013), pp. 111–14, p. 113.
66. Schmitt, 'Introduction', p. 11.
67. Hansen, *Bodies in Code*, p. 276, *n*xiii.
68. Apter, *Continental Drift*, p. 19.
69. Hansen, *Bodies in Code*, pp. 8, 147.
70. United States Holocaust Memorial Museum, 'Witnessing History', author transcript of audio text.
71. Vivian M. Patraka, *Spectacular Suffering: Theatre, Fascism and the Holocaust* (Bloomington: Indiana University Press, 1999), p. 4; Senft, 'Introduction', p. 17.
72. Redfield, 'Virtual Trauma', p. 68.
73. Sara Ahmed, *The Cultural Politics of Emotion* (New York: Routledge, 2004), pp. 8, 10.
74. United States Holocaust Memorial Museum, 'Witnessing History', author transcript of audio text.
75. Seltzer, *Serial Killers*, p. 254.
76. Redfield, 'Virtual Trauma', p. 75.
77. Senft, 'Introduction', p. 17.
78. Hansen, *Bodies in Code*, p. 5. Italics in original.

5 Affect's Spill: Theatrical 'Sensationship' in Cultures of Memory

1. Joe Kelleher and Nicholas Ridout, 'Introduction: The Spectators and the Archive' in Claudia Castellucci, Romeo Castellucci, Chiara Guidi, Joe Kelleher and Nicholas Ridout (eds.) *The Theatre of Societas Raffaello Sanzio* (London: Routledge, 2007), pp. 1–20, p. 12.
2. Margaret Hamilton, 'Art and Politics and the Zürcher Theater Spektakel: Maria Magdalena Schwaegermann talks with Margaret Hamilton', *Performance Paradigm* 3 (May 2007). http://www.performanceparadigm.net/wp-content/uploads/2007/06/08-hamilton.pdf, page accessed 24 September 2013.

3. Helena Grehan, *Performance, Ethics and Spectatorship in a Global Age* (Houndmills, Basingstoke: Palgrave Macmillan, 2009), p. 59.
4. Joe Kelleher, '*BR.#04* KunstenFESTIVALdesARTs, Brussels May 2003: Storytime' in Claudia Castellucci, Romeo Castellucci, Chiara Guidi, Joe Kelleher and Nicholas Ridout (eds.) *The Theatre of Socìetas Raffaello Sanzio* (London: Routledge, 2007), pp. 95–103, p. 100.
5. Joe Kelleher, 'The Suffering of Images' in Adrian Heathfield (ed.) *Live: Art and Performance* (New York: Routledge, 2004), pp. 192–4, p. 194.
6. Nicholas Ridout, 'Make-Believe: Socìetas Raffaello Sanzio do Theatre' in Joe Kelleher and Nicholas Ridout (eds.) *Contemporary Theatres in Europe: A Critical Companion* (London: Routledge, 2006), pp. 175–87, p. 183.
7. Romeo Castellucci cited in Grehan, *Performance, Ethics and Spectatorship*, p. 56.
8. Gabriella Calchi Novati, 'Language Under Attack: The Iconoclastic Theatre of *Socìetas Raffaello Sanzio*', *Theatre Research International* 34.1 (2009), pp. 50–65, p. 57.
9. Rebecca Schneider, *Performing Remains: Art and War in Times of Theatrical Reenactment* (Abingdon, Oxon: Routledge, 2011), p. 41. Italics in original.
10. Erika Fischer-Lichte, 'Reality and Fiction in Contemporary Theatre', *Theatre Research International* 33 (2008), pp. 84–96, p. 87.
11. Bridget Escolme, *Talking to the Audience: Shakespeare, Performance, Self* (Abingdon, Oxon: Routledge, 2005), p. 139.
12. Sam Trubridge, 'Terrifying Grief', *Performance Paradigm* 4 (May 2008), p. 15. http://www.performanceparadigm.net/wp-content/uploads/2008/06/8trubridge.pdf, page accessed 24 September 2013.
13. Ehren Fordyce, 'Whiteout' in *The Presence Project*. http://documents.stanford.edu/67/839, page accessed 12 January 2012.
14. Jill Bennett, *Empathic Vision: Affect, Trauma and Contemporary Art* (Stanford: Stanford University Press, 2005), p. 2.
15. *Ibid.*, p. 23.
16. *Ibid.*, p. 22.
17. *Ibid.*, p. 35. It is important to note that Bennett's take on affect argues for a certain complexity in affect's production of empathy by entailing (via Dominick LaCapra) forms of *'empathetic unsettlement'*, p. 8. Italics in original.
18. Jill Bennett, *Practical Aesthetics: Events, Affects and Art after 9/11* (London: I.B. Taurus, 2012), p. 3.
19. *Ibid.*, pp. 2, 6. Bennett also argues that September 11 'manifested as a phenomenological shift' whose images can be understood as co-options of 'circuits of affect; they are used, incorporated, entrained', p. 24.
20. Ann Cvetkovich and Ann Pellegrini, 'Introduction', *S&F Online* 2.1 (2003). http://sfonline.barnard.edu/ps/intro.htm, page accessed 17 September 2013.
21. Jean-Luc Nancy, *Listening*, trans. Charlotte Mandell (New York: Fordham University Press, 2002), p. 9.
22. Michael Hardt, 'What Affects Are Good For' in Patricia Ticineto Clough and Jean Halley (eds.) *The Affective Turn: Theorizing the Social* (Durham: Duke University Press, 2007), pp. ix–xiii, p. ix.

23. Joe Kelleher, 'An Organism on the Run' in Claudia Castellucci, Romeo Castellucci, Chiara Guidi, Joe Kelleher and Nicholas Ridout (eds.) *The Theatre of Socìetas Raffaello Sanzio* (London: Routledge, 2007), pp. 39–45, p. 44.
24. Nicholas Ridout, 'Welcome to the Vibratorium', *Senses and Society* 3.2 (2008), pp. 221–31, p. 223.
25. Kelleher, '*BR.#04* KunstenFESTIVALdesARTs', p. 98.
26. Lisa Saltzman, *Making Memory Matter: Strategies of Remembrance in Contemporary Art* (Chicago: University of Chicago Press, 2006), p. 13.
27. Jeanette R. Malkin, *Memory-Theater and Postmodern Drama* (Ann Arbor: University of Michigan Press, 1999). See also Karen Jürs-Munby, ' "Did You Mean *Post-Traumatic Theatre?*": The Vicissitudes of Traumatic Memory in Contemporary Postdramatic Performances', *Performance Paradigm* 5.2 (October 2009). http://www.performanceparadigm.net/wp-content/uploads/2009/10/jurs-munby-posttraumatic-postdramatic-final-copy-with-images.pdf, page accessed 24 September 2013.
28. *Ibid.*, p. 27.
29. *Ibid.*, p. 2.
30. *Ibid.*, p. 3.
31. Frances A. Yates, *The Art of Memory* (Chicago: University of Chicago Press, 2001 [1966]).
32. Malkin, *Memory-Theater*, pp. 29, 32.
33. *Ibid.*, p. 8. Italics in original.
34. Hans-Thies Lehmann, *Postdramatic Theatre*, trans. Karen Jürs-Munby (London: Routledge, 2006), p. 2.
35. Dominick LaCapra, *History and Memory After Auschwitz* (Ithaca: Cornell University Press, 1998), p. 47.
36. *Ibid.*, p. 47.
37. Romeo Castellucci interviewed by Valentina Valentini and Bonnie Marranca, 'The Universal: The Simplest Place Possible', trans. Jane House, *PAJ* 77 (2004), pp. 16–25, p. 17, p. 18.
38. Chiara Guidi and Romeo Castellucci, '10 September 2002' in Claudia Castellucci, Romeo Castellucci, Chiara Guidi, Joe Kelleher and Nicholas Ridout (eds.) *The Theatre of Socìetas Raffaello Sanzio* (London: Routledge, 2007), pp. 72–3, p. 72.
39. *Ibid.*, p. 55.
40. Grehan discusses relations between 'pre-tragic' theatre and Levinas' 'pre-ontological realm', *Performance, Ethics and Spectatorship*, p. 37.
41. Nicholas Ridout, *Stage Fright, Animals and Other Theatrical Problems* (Cambridge: Cambridge University Press, 2006), p. 117.
42. Freddy Decreus, 'The Nomadic Theatre of the *Socìetas Raffaello Sanzio*: A Case of Postdramatic Reworking of (The Classical) Tragedy' in Lorna Hardwick and Christopher Stray (eds.) *A Companion to Classical Receptions* (Massachusetts: Blackwell, 2008), p. 274–86, p. 276.
43. Ridout, *Stage Fright*, p. 117.
44. Nicholas Ridout, 'The Worst Sort of Places', *Theater* 37.3 (2007), pp. 7–15, pp. 9–10.

45. As quoted by Valentini and Marranca, 'The Universal', p. 17.
46. Kelleher and Ridout, 'Introduction', pp. 2, 3.
47. Romeo Castellucci quoted in Ridout, 'The Worst Sort of Places', p. 7.
48. Socìetas Raffaello Sanzio, 'Use of the Truth in the Sound of Scott Gibbons' in *Socìetas Raffaello Sanzio: Tragedia Endogonidia by Romeo Castellucci* (notes to the DVD), p. 77.
49. My response to this performance is documented at Bryoni Trezise, 'History's Imprints', *RealTime* 90 (April–May 2009). http://www.realtimearts.net/article/90/9395, page accessed 24 September 2013.
50. Daniel Sack, '*L.#09* – London Episode of the Tragedia Endogonidia', *Theatre Journal* 58.3 (2006), pp. 485–6, p. 486.
51. Socìetas Raffaello Sanzio, 'Use of the Truth in the Sound', p. 77.
52. Hans-Thies Lehmann, *Postdramatic Theatre* p. 166. Italics in original.
53. Ridout, 'Make-Believe', p. 177.
54. Susan Sontag, *Regarding the Pain of Others* (London: Hamish Hamilton, 2003), pp. 36, 16.
55. *Ibid.*, pp. 19, 64.
56. Nicholas Ridout, *Theatre & Ethics* (Houndmills, Basingstoke: Palgrave Macmillan, 2009), p. 67.
57. Elaine Scarry, *The Body in Pain: The Making and Unmaking of the World* (New York: Oxford University Press, 1985), p. 4.
58. See, for instance, Naomi Mandel, 'Rethinking "After Auschwitz": Against a Rhetoric of the Unspeakable in Holocaust Writing', *Boundary 2* 28.2 (2001), pp. 203–28.
59. I would like to acknowledge that this idea was initially Caroline Wake's, as noted in Bryoni Trezise and Caroline Wake, 'Introduction to After Effects: Performing the Ends of Memory', *Performance Paradigm* 5.1 (May 2009). http://www.performanceparadigm.net/wp-content/uploads/2009/07/wake-and-trezise-intro-final.pdf, page accessed 24 September 2013.
60. See Diana Taylor, *The Archive and the Repertoire: Performing Cultural Memory in the Americas* (Durham: Duke University Press, 2003); Susan Leigh Foster, *Choreographing Empathy: Kinesthesia in Performance* (Abingdon, Oxon: Routledge, 2011); Joseph Roach, *Cities of the Dead: Circum-Atlantic Performance* (New York: Columbia University Press, 1996).
61. Foster, *Choreographing Empathy*, p. 127.
62. Gregory J. Seigworth and Melissa Gregg, 'An Inventory of Shimmers' in Melissa Gregg and Gregory J. Seigworth (eds.) *The Affect Theory Reader* (Durham: Duke University Press, 2010), pp. 1–28, p. 1. Italics in original.
63. Denise Varney, 'Gestus, Affect and the Post-Semiotic in Contemporary Theatre', *The International Journal of the Arts in Society* 1.3 (2007), pp. 113–20, p. 116.
64. *Ibid.*, p. 114.
65. Bennett, *Empathic Vision*, p. 10.
66. *Ibid.*, p. 10. Italics in original.
67. Ahmed, *The Cultural Politics of Emotion* (New York: Routledge, 2004), pp. 30, 10.
68. Nancy, *Listening*, p. 9.

69. Greg Urban, 'Ritual Wailing in Amerindian Brazil', *American Anthropologist* 90.2 (June 1988), pp. 385–400, p. 386.
70. Russell-Horschild quoted by Helga Kotthoff in 'Affect and Meta-Affect in Georgian Mourning Rituals' in Jürgen Schlaeger and Gesa Stedman (eds.) *Representations of Emotions* (Tübingen: Gunter Narr Verlag, 1999), pp. 149–72, p. 152.
71. *Ibid.*, p. 153.
72. Anna Caraveli quoted by Kotthoff, 'Affect and Meta-Affect in Georgian Mourning Rituals', p. 150.
73. Kotthoff, 'Affect and Meta-Affect in Georgian Mourning Rituals', p. 151.
74. Nancy, *Listening*, p. 6.
75. *Ibid.*, p. 10.
76. Erin Manning, *Politics of Touch: Sense, Movement, Sovereignty* (Minneapolis: University of Minnesota Press, 2007), p. xv.
77. I discuss Mark Hansen's work in detail in Chapter 4 of this book. Mark B.N. Hansen, *Bodies in Code: Interfaces with Digital Media* (New York: Routledge, 2006).
78. Laura Cull, 'Performance as Philosophy: Responding to the Problem of "Application"', *Theatre Research International* 37.1 (March 2012), pp. 20–7, p. 24. Italics in original.

Bibliography

Ahmed, Sara. *Strange Encounters: Embodied Others in Post-Coloniality* (London: Routledge, 2000).

Ahmed, Sara. *The Cultural Politics of Emotion* (New York: Routledge, 2004).

Ahmed, Sara. 'The Non-Performativity of Anti-Racism', *Borderlands e-journal* 5.3 (2005). http://www.borderlands.net.au/vol5no3_2006/ahmed_non perform.htm, page accessed 30 August 2013.

Ahmed, Sara. 'Happy Objects' in Melissa Gregg and Gregory J. Seigworth (eds.) *The Affect Theory Reader* (Durham: Duke University Press, 2010), pp. 29–51.

Anne Frank Museum. 'The Secret Annex Online'. http://www.annefrank.org/en/Subsites/Home/, page accessed 17 September 2013.

Apel, Dora. *Memory Effects: The Holocaust and the Art of Secondary Witnessing* (New Brunswick: Rutgers, 2002).

Apter, Emily. *Continental Drift: From National Characters to Virtual Subjects* (Chicago: University of Chicago Press, 1999).

Arena TV Australia. *Crossing Over* (November 2005).

Arthur, Paul Longley. 'Exhibiting History: The Digital Future', *reCollections: Journal of the National Museum of Australia* 3.1 (2008), pp. 33–50.

Assmann, Jan. 'Collective Memory and Cultural Identity', trans. John Czaplicka *New German Critique* 65 (1995), pp. 125–33.

Auschwitz-Birkenau Memorial and Museum. 'Virtual Tour of Auschwitz Sites'. http://en.auschwitz.org/z/index.php?option=com_content&task=view&id=6&Itemid=8, page accessed 17 September 2013.

Bain, Andrew. 'Destination Dreaming', *Australian Geographic* 88 (October–December 2007), pp. 72–85.

Bartov, Omer. 'Defining Enemies, Making Victims: Germans, Jews, and the Holocaust', *The American Historical Review* 103.3 (1998), pp. 771–816.

BBC World News, 'Berlin Opens Holocaust Memorial', *BBC World News* (10 May 2005). http://news.bbc.co.uk/2/hi/4531669.stm, page accessed 13 September 2013.

Bennett, Jill. *Empathic Vision: Affect, Trauma and Contemporary Art* (Stanford: Stanford University Press, 2005).

Bennett, Jill. *Practical Aesthetics: Events, Affects and Art after 9/11* (London: I.B. Taurus, 2012).

Berger, James. 'Which Prosthetic? Mass Media, Narrative, Empathy, and Progressive Politics', *Rethinking History* 11.4 (2007), pp. 597–612.

Berlant, Lauren. 'The Subject of True Feeling: Pain, Privacy, and Politics' in Karyn Ball (ed.) *Traumatizing Theory: The Cultural Politics of Affect in and Beyond Psychoanalysis* (New York: Other Press, 2007), pp. 305–47.

Best, Susan. 'Rethinking Visual Pleasure: Aesthetics and Affect', *Theory Psychology* 17.4 (2007), pp. 505–14.

Bishop, Claire. *Artificial Hells: Participatory Art and the Politics of Spectatorship* (London: Verso, 2012).
Bolter, Jay David and Richard Grusin. *Remediation: Understanding New Media* (Massachusetts: MIT Press, 2000 [1999]).
Brennan, Teresa. *The Transmission of Affect* (Ithaca, New York: Cornell University Press, 2004).
Burns, Alex and Ben Eltham. 'Twitter Free Iran: An Evaluation of Twitter's Role in Public Diplomacy and Information Operations in Iran's 2009 Election Crisis', *Communications Policy & Research Forum*, 19–20 November 2009, University of Technology, Sydney. http://eprints.vu.edu.au/15230/, page accessed 13 September 2013.
Butler, Judith. 'Afterword: After Loss, What Then?' in David L. Eng and David Kazanjian (eds.) *Loss: The Politics of Mourning* (Berkeley: University of California Press, 2003), pp. 467–73.
Butler, Judith. *Precarious Life: The Powers of Mourning and Violence* (London: Verso, 2004).
Butler, Kelly Jean. *Witnessing Australian Stories: History, Testimony, and Memory in Contemporary Culture* (New Jersey: Transaction Publishers, 2013).
Callard, Felicity and Constantina Papoulias. 'Affect and Embodiment' in Susannah Radstone and Bill Schwarz (eds.) *Memory: Histories, Theories, Debates* (New York: Fordham University Press, 2010), pp. 246–62.
Caruth, Cathy. *Unclaimed Experience: Trauma, Narrative, and History* (Baltimore: The Johns Hopkins University Press, 1996).
Casey, Edward S. 'Habitual Body and Memory in Merleau-Ponty', *Man and World* 17 (1984), pp. 279–97.
Casey, Edward S. *Remembering: A Phenomenological Study*, 2nd edition (Bloomington: Indiana University Press, 2000 [1987]).
Casey, Maryrose. 'Carnivalising Sovereignty: Containing the Indigenous Protest within the "White" Australian Nation', *About Performance* 7 (2007), pp. 69–84.
Christopher, Kevin. ' "I Speak to Dead People": Medium John Edward Hosts SciFi Cable Show', *Skeptical Inquirer* 24.5 (2000), p. 9.
Clark, Laurie Beth. 'Always Already Again: Trauma Tourism and the Politics of Memory Culture', *Encounters* 1 (2010), pp. 65–74.
Clark, Mary Marshall. 'The September 11, 2001, Oral History Narrative and Memory Project: A First Report', *The Journal of American History* 89.2 (September 2002), pp. 569–79.
Clough, Patricia Ticineto. 'Introduction' in Patricia Ticineto Clough and Jean Halley (eds.) *The Affective Turn: Theorizing the Social* (Durham: Duke University Press, 2007), pp. 1–33.
Cohen, Michael, Paul Dwyer and Laura Ginters. 'Performing "Sorry Business": Reconciliation and Redressive Action' in Graham St John (ed.) *Victor Turner and Contemporary Cultural Performance* (Oxford: Berghahn Books, 2008), pp. 76–93.
Connerton, Paul. *The Spirit of Mourning: History, Memory and the Body* (Cambridge: Cambridge University Press, 2011).

Cowlishaw, Gillian. 'Mythologising Culture Part 1: Desiring Aboriginality in the Suburbs', *Australian Journal of Anthropology* 21 (2010), pp. 208-27.

Crownshaw, Richard. *The Afterlife of Holocaust Memory in Contemporary Literature and Culture* (Houndmills, Bassingstoke: Palgrave, 2010).

Csordas, Thomas J. 'Somatic Modes of Attention', *Cultural Anthropology* 8.2 (1993), pp. 135-56.

Cull, Laura. 'Performance as Philosophy: Responding to the Problem of "Application"', *Theatre Research International* 37.1 (March 2012), pp. 20-7.

Cvetkovich, Ann and Ann Pellegrini. 'Introduction', *S&F Online* 2.1 (2003). http://sfonline.barnard.edu/ps/intro.htm, page accessed 17 September 2013.

Dachau Concentration Camp Memorial Site. 'Virtual Tour', *Dachau Concentration Camp Memorial Site*. http://www.kz-gedenkstaette-dachau.de/virtual_tour.html, page accessed 13 September 2013.

Daily Telegraph, The. 'Outrage Over Melbourne Artist Jane Korman's I Will Survive Dance at Polish Death Camp', *The Daily Telegraph* (14 July 2010). http://www.theaustralian.com.au/news/outrage-over-melbourne-artist-jane-kormans-i-will-survive-dance-at-polish-death-camp/story-e6frg6of-1225891392172, page accessed 13 September 2013.

Damasio, Antonio R. *The Feeling of What Happens: Body and Emotion in the Making of Conciousness* (New York: Harcourt Brace and Co., 1999).

Dapin, Mark. 'Lest We Remember', *The Sydney Morning Herald Good Weekend* (8 July 2006), pp. 33-4.

Davis, Erik. *TechGnosis: Myth, Magic & Mysticism in the Age of Information* (New York: Three Rivers Press, 1998).

Decreus, Freddy. 'The Nomadic Theatre of the *Societas Raffaello Sanzio*: A Case of Postdramatic Reworking of (the Classical) Tragedy' in Lorna Hardwick and Christopher Stray (eds.) *A Companion to Classical Receptions* (Massachusetts: Blackwell, 2008), p. 274-86.

Dekel, Irit. 'Ways of Looking: Observation and Transformation at the Holocaust Memorial, Berlin', *Memory Studies* 2.1 (2009), pp. 71-86.

Dibbell, Julian. 'A Rape in Cyberspace', *The Village Voice* (21 December 1993), pp. 36-42.

Dimock, Wai Chee. 'Introduction: Genres as Fields of Knowledge', *PMLA* 122.5 (2007), pp. 1377-88.

Doná, Giorgia. 'Collective Suffering and Cyber-Memorialisation in Post-Genocide Rwanda' in Mick Broderick and Antonio Traverso (eds.) *Trauma, Media, Art* (Newcastle Upon Tyne: Cambridge Scholars Publishing, 2010), pp. 16-35.

Douglas, Kate. 'Cyber-Commemoration: Life Writing, Trauma and Memorialisation', *Life Writing Symposium* 13-15 June 2006, Flinders University.

Doss, Erika. *Memorial Mania: Public Feeling in America* (Chicago: University of Chicago Press, 2010).

Dovey, Jon. *Freakshow: First Person Media and Factual Television* (London: Pluto Press, 2000).

Dyer, Pam, Lucinda Aberdeen and Sigrid Schuler. 'Tourism Impacts on an Australian Indigenous Community: A Djapugay Case Study', *Tourism Management* 24 (2004), pp. 83–95.
Edward, John. *Crossing Over: The Stories Behind the Stories* (New York: Princess Books, 2001).
Elder, Catriona, Angela Pratt and Cath Ellis. 'Running Race: Reconciliation, Nationalism and the Sydney 2000 Olympic Games', *International Review for the Sociology of Sport* 41.2 (2006) pp. 181–200.
Ellison, David A. 'The Spoiler's Art: Embarrassed Space as Memorialisation', *The South Atlantic Quarterly* 110.1 (2011), pp. 89–100.
Erll, Astrid and Ann Rigney. 'Introduction: Cultural Memory and Its Dynamics' in Astrid Erll and Ann Rigney (eds.) *Mediation, Remediation and the Dynamics of Cultural Memory* (Berlin: Walter de Gruyter, 2009), pp. 1–14.
Escolme, Bridget. *Talking to the Audience: Shakespeare, Performance, Self* (Abingdon, Oxon: Routledge, 2005).
Fischer-Lichte, Erika. 'Reality and Fiction in Contemporary Theatre', *Theatre Research International* 33 (2008), pp. 84–96.
Fisher, Jennifer. 'Relational Sense: Towards a Haptic Aesthetics', *Parachute* 87 (July–September, 1997), pp. 4–11.
Fordyce, Ehren. 'Whiteout' in *The Presence Project*. http://documents.stanford.edu/67/839, page accessed 12 January 2012.
Foster, Susan Leigh. *Choreographing Empathy: Kinesthesia in Performance* (Abingdon, Oxon: Routledge, 2011).
Garde-Hansen, Joanne, Andrew Hoskins and Anna Reading, 'Introduction' in Joanne Garde-Hansen, Andrew Hoskins and Anna Reading (eds.) *Save As...Digital Memories* (Houndmills, Basingstoke: Palgrave Macmillan, 2009), pp. 1–21.
Gibbs, Anna. 'Disaffected', *Continuum* 16.3 (2002), pp. 335–41.
Gliatto, Tom and Natasha Stoynoff. 'Medium Rare', *People Weekly* 57.17 (6 May 2002), pp. 85–6.
Godfrey, Mark. *Abstraction and the Holocaust* (New Haven: Yale University Press, 2007).
Goodall, Jane R. *Performance and Evolution in the Age of Darwin: Out of the Natural Order* (London: Routledge, 2002).
Gordon, Avery F. *Ghostly Matters: Haunting and the Sociological Imagination* (Minneapolis: University of Minnesota Press, 1997).
Grehan, Helena. *Performance, Ethics and Spectatorship in a Global Age* (Houndmills, Basingstoke: Palgrave Macmillan, 2009).
Grindstaff, Laura. *The Money Shot: Trash, Class and the Making of TV Talk Shows* (Chicago: University of Chicago Press, 2002).
Grusin, Richard. 'Remediation and Premediation', *Criticism* 46.1 (2004), pp. 17–40.
Guidi, Chiara and Romeo Castellucci. '10 September 2002' in Claudia Castellucci, Romeo Castellucci, Chiara Guidi, Joe Kelleher and Nicholas Ridout (eds.) *The Theatre of Societas Raffaello Sanzio* (London: Routledge, 2007), pp. 72–3.

Gunew, Sneja. *Haunted Nations: The Colonial Dimensions of Multiculturalisms* (Abingdon, Oxon: Routledge, 2004).
Haaretz News Service, 'Jewish Artist Defends YouTube Video "Dancing Auschwitz"', *Haaretz News Service* (10 July 2010). http://www.haaretz.com/jewish-world/jewish-artist-defends-youtube-video-dancing-auschwitz-1.301096, page accessed 13 September 2013.
Hage, Ghassan. *White Nation: Fantasies of White Supremacy in a Multicultural Society* (New York: Routledge, 2000).
Hage, Ghassan. *Against Paranoid Nationalism: Searching for Hope in a Shrinking Society* (Sydney: Pluto Press, 2003).
Halbwachs, Maurice. *On Collective Memory*, trans. Lewis A. Coser (Chicago: University of Chicago Press, 1992 [1941]).
Hamilton, Margaret. 'Art and Politics and the Zürcher Theater Spektakel: Maria Magdalena Schwaegermann Talks with Margaret Hamilton', *Performance Paradigm* 3 (May 2007). http://www.performanceparadigm.net/wp-content/uploads/2007/06/08-hamilton.pdf, page accessed 24 September 2013.
Hansen, Mark B. N. *Bodies in Code: Interfaces with Digital Media* (New York: Routledge, 2006).
Hardt, Michael. 'What Affects Are Good For' in Patricia Ticineto Clough and Jean Halley (eds.) *The Affective Turn: Theorizing the Social* (Durham: Duke University Press, 2007), pp. ix–xiii.
Harris, Geraldine. 'The Ethics and Politics of Witnessing Whoopi' in Bryoni Trezise and Caroline Wake (eds.) *Visions and Revisions: Performance, Memory, Trauma* (Copenhagen: Museum Tusculanum Press, 2013), pp. 115–30.
Healy, Chris. *Forgetting Aborigines* (Sydney: UNSW Press, 2008).
Hemmings, Clare. 'Invoking Affect', *Cultural Studies* 19.5 (2005), pp. 548–67.
Henry, Rosita. 'Dancing into Being: The Tjapukai Aboriginal Cultural Park and the Laura Dance Festival', *Australian Journal of Anthropology* 11.3 (2000), pp. 322–32.
Highmore, Ben. 'Bitter After Taste: Affect, Food, and Social Aesthetics' in Melissa Gregg and Gregory J. Seigworth (eds.) *The Affect Theory Reader* (Durham: Duke University Press, 2010), pp. 118–37.
Hill, John. *A Daily Dose of Architecture* (11 May 2005). http://www.archidose.blogspot.com/2005/05/stelae.html, page accessed 13 September 2013.
Hirsch, Marianne. *Family Frames: Photography, Narrative and Postmemory* (Cambridge, Massachusetts: Harvard University Press, 1997).
Hirsch, Marianne. 'Surviving Images: Holocaust Photographs and the Work of Postmemory', *Yale Journal of Criticism* 14.1 (2001), pp. 5–37.
Hirsch, Marianne. 'The Generation of Postmemory', *Poetics Today* 29.1 (2008), pp. 103–28.
Hollander, Jason. 'Peter Eisenman, Architecture '60, Designs New Holocaust Memorial in Berlin', *Columbia News* (10 July 2003). http://www.columbia.edu/cu/news/03/07/peterEisenman.html, page accessed 13 September 2013.
Hoskins, Andrew. 'Anachronisms of Media, Anachronisms of Memory: From Collective Memory to a New Memory Ecology' in Motti Neiger, Oren Meyers

and Eyal Zandberg (eds.) *On Media Memory: Collective Memory in a New Media Age* (New York: Palgrave Macmillan, 2011), pp. 278–88.
Hrvatin, Emil. 'The Scream', *Performance Research* 2.1 (1997), pp. 82–91.
Huyssen, Andreas. *Twilight Memories: Marking Time in a Culture of Amnesia* (New York: Routledge, 1995).
Immordino-Yang, Mary Helen and Antonio Damasio. 'We Feel, Therefore We Learn: The Relevance of Affective and Social Neuroscience to Education', *Mind, Brain and Education* 1.1 (2007), pp. 3–10.
Israeli Center for Digital Art, The. Artur Żmijewski *80064* (2005), *The Israeli Center for Digital Art*. http://www.digitalartlab.org.il/ArchiveVideo.asp?id=16, page accessed 8 May 2013.
Jameson, Fredric. *Postmodernism, or, the Cultural Logic of Late Capitalism* (Durham: Duke University Press, 2001 [1991]).
Jaroff, Leon. 'Talking to the Dead', *Time* 157.9 (5 March 2001), p. 52.
Jewish Museum Berlin, The. 'The Installations', *The Jewish Museum Berlin*. http://www.jmberlin.de/main/EN/01-Exhibitions/04-installations.php, page accessed 14 September 2013.
Jewish Museum Berlin, 'What We Won't Show You', *The Jewish Museum Berlin*. http://www.jmberlin.de/osk/wwnz/filme/en/film-7/film.php, page accessed 17 September 2013.
Jürs-Munby, Karen. ' "Did You Mean *Post-Traumatic Theatre?*": The Vicissitudes of Traumatic Memory in Contemporary Postdramatic Performances', *Performance Paradigm* 5.2 (October 2009). http://www.performanceparadigm.net/wp-content/uploads/2009/10/jurs-munby-posttraumatic-postdramatic-final-copy-with-images.pdf, page accessed 24 September 2013.
Kansteiner, Wulf. 'Finding Meaning in Memory: A Methodological Critique of Collective Memory Studies', *History and Theory* 41 (May 2002), pp. 179–97.
Kaplan, E. Ann. *Trauma Culture: The Politics of Terror and Loss in Media and Literature* (New Brunswick: Rutgers, 2005).
Kelleher, Joe. 'The Suffering of Images' in Adrian Heathfield (ed.) *Live: Art and Performance* (New York: Routledge, 2004), pp. 192–4.
Kelleher, Joe. 'An Organism on the Run' in Claudia Castellucci, Romeo Castellucci, Chiara Guidi, Joe Kelleher and Nicholas Ridout (eds.) *The Theatre of Societas Raffaello Sanzio* (London: Routledge, 2007), pp. 39–45.
Kelleher, Joe. '*BR.#04* KunstenFESTIVALdesARTs, Brussels May 2003: Storytime' in Claudia Castellucci, Romeo Castellucci, Chiara Guidi, Joe Kelleher and Nicholas Ridout (eds.) *The Theatre of Societas Raffaello Sanzio* (London: Routledge, 2007), pp. 95–103.
Kelleher, Joe and Nicholas Ridout. 'Introduction: The Spectators and the Archive' in Claudia Castellucci, Romeo Castellucci, Chiara Guidi, Joe Kelleher and Nicholas Ridout (eds.) *The Theatre of Societas Raffaello Sanzio* (London: Routledge, 2007), pp. 1–20.
Kirby, Vicki. *Telling Flesh: The Substance of the Corporeal* (New York: Routledge, 1997).
Kirshenblatt-Gimblett, Barbara. *Destination Culture: Tourism, Museums, and Heritage* (Berkeley: University of California Press, 1998).

Korman, Jane. *Jane Korman Art.* http://www.janekormanart.com/janekormanart.com/DA_Hype.html, page accessed 13 September 2013.
Kosofsky Sedgwick, Eve. *Touching Feeling: Affect, Pedagogy, Performativity* (Durham: Duke University Press, 2003).
Kotthoff, Helga. 'Affect and Meta-Affect in Georgian Mourning Rituals' in Jürgen Schlaeger and Gesa Stedman (eds.) *Representations of Emotions* (Tübingen: Gunter Narr Verlag, 1999), pp. 149–72.
LaCapra, Dominick. *History and Memory After Auschwitz* (Ithaca: Cornell University Press, 1998).
Landsberg, Alison. *Prosthetic Memory: The Transformation of American Remembrance in the Age of Mass Culture* (New York: Columbia University Press, 2004).
Landsberg, Alison. 'Memory, Empathy, and the Politics of Identification', *International Journal of Politics, Culture and Society* 22 (2009), pp. 221–9.
Langton, Marcia. 'The Edge of the Sacred, the Edge of Death: Sensual Inscriptions' in Bruno David and Meredith Wilson (eds.) *Inscribed Landscapes: Marking and Making Place* (Honolulu, HI: University Of Hawai'i Press, 2002), pp. 253–69.
Langton, Marcia. 'Earth, Wind, Fire, and Water: The Social and Spiritual Construction of Water in Aboriginal Societies' in Bruno David, Bryce Barker and Ian J. McNiven (eds.) *The Social Archaeology of Australian Indigenous Societies* (Canberra: Aboriginal Studies Press, 2006), pp. 139–59.
Lehmann, Hans-Thies. *Postdramatic Theatre*, trans. Karen Jürs-Munby (London: Routledge, 2006).
Malkin, Jeanette R. *Memory-Theater and Postmodern Drama* (Ann Arbor: University of Michigan Press, 1999).
Mandel, Naomi. 'Rethinking "After Auschwitz": Against a Rhetoric of the Unspeakable in Holocaust Writing', *Boundary 2* 28.2 (2001), pp. 203–28.
Manning, Erin. *Politics of Touch: Sense, Movement, Sovereignty* (Minneapolis: University of Minnesota Press, 2007).
Marker, Chris. *Sans Soleil* (1983) Argos Films, France.
Massumi, Brian. *Parables for the Virtual: Movement, Affect, Sensation* (Durham: Duke University Press, 2002).
Massumi, Brian. 'The Future Birth of the Affective Fact: The Political Ontology of Threat' in Melissa Gregg and Gregory J. Seigworth (eds.) *The Affect Theory Reader* (Durham: Duke University Press, 2010), pp. 52–70.
Mortensen, Mette. 'When Citizen Photojournalism Sets the News Agenda: Neda Agha Soltan as a Web 2.0 Icon of Post-Election Unrest in Iran', *Global Media and Communication* 7.1 (2011), pp. 4–16.
Muecke, Stephen. 'Lonely Representations: Aboriginality and Cultural Studies', *Journal of Australian Studies* 16.35 (1992), pp. 32–44.
Muecke, Stephen. *Ancient and Modern: Time, Culture and Indigenous Philosophy* (Sydney: UNSW Press, 2004).
Nancy, Jean-Luc. *Listening*, trans. Charlotte Mandell (New York: Fordham University Press, 2002).

Neistat, Aimee. 'Dancing on the Ashes', *The Jerusalem Post.com* (10 August 2010). http://www.jpost.com/Magazine/Features/Article.aspx?id=190503, page accessed 13 September 2013.
Nora, Pierre. 'Between Memory and History: *Les Lieux de Mémoire*', trans. Marc Roudebush *Representations* 26 (Spring 1989), pp. 7–25.
Novati, Gabriella Calchi. 'Language Under Attack: The Iconoclastic Theatre of *Societas Raffaello Sanzio*', *Theatre Research International* 34.1 (2009), pp. 50–65.
Parsons, Michael. 'The Tourist Corroboree in South Australia to 1911', *Aboriginal History* 21 (1997), pp. 46–69.
Parsons, Michael. ' "Ah that I Could Convey a Proper Idea of this Interesting Wild Play of the Natives": Corroborees and the Rise of Indigenous Australian Cultural Tourism', *Australian Aboriginal Studies* 2 (2002), pp. 14–26.
Patraka, Vivian M. 'Situating History and Difference: The Performance of the Term *Holocaust* in Public Discourse' in Jonathan Boyarin and Daniel Boyarin (eds.) *Jews and Other Differences: The New Jewish Cultural Studies* (Minneapolis: University of Minnesota Press, 1997), pp. 54–78.
Patraka, Vivian M. *Spectacular Suffering: Theatre, Fascism and the Holocaust* (Bloomington: Indiana University Press, 1999).
Perera, Suvendrini and Joseph Pugliese. ' "Racial Suicide": The Re-licensing of Racism in Australia', *Race Class* 39.1 (1997), pp. 1–19.
Phelan, Peggy. *Unmarked: The Politics of Performance* (London: Routledge, 1993).
Phelan, Peggy. *Mourning Sex: Performing Public Memories* (New York: Routledge, 1997).
Poignant, Roslyn. *Professional Savages: Captive Lives and Western Spectacle* (New Haven: Yale University Press, 2004).
Potts, John. 'The Idea of the Ghost' in John Potts and Edward Scheer (eds.) *Technologies of Magic: A Cultural Study of Ghosts, Machines and the Uncanny* (Sydney: Power Publications, 2006), pp. 78–91.
Povinelli, Elizabeth A. *The Cunning of Recognition: Indigenous Alterities and the Making of Australian Multiculturalism* (Durham: Duke University Press, 2002).
Pruchnic, Jeff and Kim Lacey. 'The Future of Forgetting: Rhetoric, Memory, Affect', *Rhetoric Society Quarterly* 41.5 (2011), pp. 472–94.
Radstone, Susannah and Katherine Hodgkin, 'Believing the Body' in Susannah Radstone and Katherine Hodgkin (eds.) *Regimes of Memory* (London: Routledge, 2003), pp. 23–6.
Radstone, Susannah and Katharine Hodgkin (eds.) *Memory Cultures: Memory, Subjectivity and Recognition* (New Brunswick: Transaction, 2009).
Reading, Anna. 'Memory and Digital Media: Six Dynamics of the Globital Memory Field' in Motti Neiger, Oren Meyers and Eyal Zandberg (eds.) *On Media Memory: Collective Memory in a New Media Age* (New York: Palgrave Macmillan, 2011), pp. 241–52.
Redfield, Marc. 'Virtual Trauma: The Idiom of 9/11', *Diacritics* 37.1 (2007), pp. 55–80.

Ridout, Nicholas. 'Make-Believe: Socìetas Raffaello Sanzio do Theatre' in Joe Kelleher and Nicholas Ridout (eds.) *Contemporary Theatres in Europe: A Critical Companion* (London: Routledge, 2006), pp. 175–87.
Ridout, Nicholas. *Stage Fright, Animals and Other Theatrical Problems* (Cambridge: Cambridge University Press, 2006).
Ridout, Nicholas. 'The Worst Sort of Places', *Theater* 37.3 (2007), pp. 7–15.
Ridout, Nicholas. 'Welcome to the Vibratorium', *Senses and Society* 3.2 (2008), pp. 221–31.
Ridout, Nicholas. *Theatre & Ethics* (Houndmills, Basingstoke: Palgrave Macmillan, 2009).
Roach, Joseph. *Cities of the Dead: Circum-Atlantic Performance* (New York: Columbia University Press, 1996).
Rowe, David and Deborah Stevenson. 'Sydney 2000: Sociality and Spatiality in Global Media Events' in Alan Tomlinson and Christopher Young (eds.) *National Identity and Global Sports Events: Culture, Politics, and Spectacle in the Olympics and the Football World Cup* (Albany: State University of New York Press, 2006), pp. 197–214.
Rudoren, Jodi. 'Proudly Bearing Elders' Scars, Their Skin Says "Never Forget"', *New York Times* New York Edition (1 October 2012), p. A1.
Ryan, Chris and Jeremy Huyton. 'Tourists and Aboriginal People', *Annals of Tourism Research* 29.3, pp. 631–47.
Sack, Daniel. '*L.#09* — London Episode of the Tragedia Endogonidia', *Theatre Journal* 58.3 (2006), pp. 485–6.
Saltzman, Lisa. *Making Memory Matter: Strategies of Remembrance in Contemporary Art* (Chicago: University of Chicago Press, 2006).
Sampson, Steven L. 'From Reconciliation to Coexistence', *Public Culture* 15.1 (2003), pp. 181–6.
Scarry, Elaine. *The Body in Pain: The Making and Unmaking of the World* (New York: Oxford University Press, 1985).
Schlör, Joachim. *Memorial to the Murdered Jews in Europe*, trans. Paul Aston (Munich: Prestel, 2005).
Schmitt, Cannon. 'Introduction: Materia Media', *Criticism* 46.1 (2004), pp. 11–15.
Schneider, Rebecca. *Performing Remains: Art and War in Times of Theatrical Reenactment* (Abingdon, Oxon: Routledge, 2011).
Sconce, Jeffrey. *Haunted Media: Electronic Presence from Telegraphy to Television* (Durham: Duke University Press, 2000).
Seigworth, Gregory J. and Melissa Gregg. 'An Inventory of Shimmers' in Melissa Gregg and Gregory J. Seigworth (eds.) *The Affect Theory Reader* (Durham: Duke University Press, 2010), pp. 1–28.
Seiler, Joey. 'Holocaust Museum Launching Kristallnacht Second Life Exhibit with Involve', *Engage Digital* (5 December 2008). www.engagedigital.com/2008/12/05/holocaust-museum-launching-kristallnacht-second-life-exhibit-with-involve/, page accessed 18 September 2013.

Seltzer, Mark. *Serial Killers: Death and Life in America's Wound Culture* (New York: Routledge, 1998).
Senft, Theresa, M. 'Introduction: Performing the Digital Body — A Ghost Story', *Women & Performance: A Journal of Feminist Theory* 9.1 (1996), pp. 9–33.
Seremetakis, C. Nadia. 'The Memory of the Senses, Part 1: Marks of the Transitory' in C. Nadia Seremetakis (ed.) *The Senses Still: Perception and Memory as Material Culture in Modernity* (Chicago: University of Chicago Press, 1994), pp. 1–18.
Shouse, Eric. 'Feeling, Emotion, Affect', *M/C Journal* 8.6 (2005), para. 15. http://www.journal.media-culture.org.au/0512/03-shouse.php, page accessed 10 September 2013.
Silverman, Kaja. *The Threshold of the Visible World* (New York: Routledge, 1996).
Simon, Joan. *Ann Hamilton: An Inventory of Objects* (New York: Gregory R. Miller and Co., 2006).
Socìetas Raffaello Sanzio, 'Use of the Truth in the Sound of Scott Gibbons' in *Socìetas Raffaello Sanzio: Tragedia Endogonidia by Romeo Castellucci* (notes to the DVD), p. 77.
Solomon-Godeau, Abigail. 'Mourning or Melancholia: Christian Boltanski's "Missing House"', *Oxford Art Journal* 21.2 (1998), pp. 3–20.
Sontag, Susan. *Regarding the Pain of Others* (London: Hamish Hamilton, 2003).
Stevens, Quentin. 'Nothing More than Feelings: Abstract Memorials', *Architectural Theory Review* 14.2 (2009), pp. 156–72.
Stoller, Paul. *Embodying Colonial Memories: Spirit Possession, Power, and the Hauka in West Africa* (New York: Routledge, 1995).
Sutton, John, Celia B. Harris and Amanda J. Barnier. 'Memory and Cognition' in Susannah Radstone and Bill Schwarz (eds.) *Memory: Histories, Theories, Debates* (New York: Fordham University Press, 2010), pp. 209–26.
Taylor, Diana. *The Archive and the Repertoire: Performing Cultural Memory in the Americas* (Durham: Duke University Press, 2003).
Thrift, Nigel. *Non-Representational Theory: Space, Politics, Affect* (Hoboken: Routledge, 2007).
Titchener, E.B. 'Affective Memory', *The Philosophical Review* 4. 1 (Jan 1895), pp. 65–76.
Tjapukai Aboriginal Cultural Park. 'Tjapukai 25 Years On' and 'Night Tour' http://www.tjapukai.com.au/, page accessed 15 September 2013.
Trezise, Bryoni. 'History's Imprints', *RealTime* 90 (April–May 2009). http://www.realtimearts.net/article/90/9395, page accessed 24 September 2013.
Trezise, Bryoni and Caroline Wake. 'Introduction to After Effects: Performing the Ends of Memory', *Performance Paradigm* 5.1 (May 2009). http://www.performanceparadigm.net/wp-content/uploads/2009/07/wake-and-trezise-intro-final.pdf, page accessed 13 September 2013.

Trubridge, Sam. 'Terrifying Grief', *Performance Paradigm* 4 (May 2008), p. 15. http://www.performanceparadigm.net/wp-content/uploads/2008/06/8trubridge.pdf, page accessed 24 September 2013.

Tumarkin, Maria. *Traumascapes: The Power and Fate of Places Transformed by Tragedy* (Carlton: Melbourne University Press, 2005).

United States Holocaust Memorial Museum. 'Witnessing History: Kristallnacht— The November 1938 Pogroms'. http://snurl.com/7r1i6, page accessed 17 September 2013.

University of Southern California Shoah Foundation. 'IWitness'. http://iwitness.usc.edu/SFI/, page accessed 17 September 2013.

Urban, Greg. 'Ritual Wailing in Amerindian Brazil', *American Anthropologist* 90.2 (June 1988), pp. 385–400.

Urban, Richard, Paul Marty and Michael Twidale. 'A Second Life for Your Museum: 3D Multi-User Virtual Environments and Museums' in *Museums and the Web 2007: The International Conference for Culture and Heritage On-line*. 11–14 April, 2007, San Francisco, California. http://www.museumsandtheweb.com/mw2007//papers/urban/urban.html, page accessed 18 September 2013.

Valentini, Valentina and Bonnie Marranca. 'The Universal: The Simplest Place Possible' trans. Jane House, *PAJ* 77 (2004), pp. 16–25.

Van Dijck, Jose. *Mediated Memories in the Digital Age* (Stanford: Stanford University Press, 2007).

Varney, Denise. 'Gestus, Affect and the Post-Semiotic in Contemporary Theatre', *The International Journal of the Arts in Society* 1.3 (2007), pp. 113–20.

Wake, Caroline. 'Regarding the Recording: The Viewer of Video Testimony, the Complexity of Copresence and the Possibility of Tertiary Witnessing', *History and Memory* 25.1 (2013), pp. 111–14.

Wake, Caroline. 'The Accident and the Account: Towards a Taxonomy of Spectatorial Witness in Theatre and Performance Studies' in Bryoni Trezise and Caroline Wake (eds.) *Visions and Revisions: Performance, Memory, Trauma* (Copenhagen: Museum Tusculanum Press, 2013), pp. 33–56.

Weissman, Gary. *Fantasies of Witnessing: Postwar Efforts to Experience the Holocaust* (Ithaca: Cornell University Press, 2004).

Werrett, Simon. *Fireworks: Pyrotechnic Arts and Sciences in European History* (Chicago: University of Chicago Press, 2010).

Whitlock, Gillian. 'Remediating Gorilla Girl: Rape Warfare and the Limits of Humanitarian Storytelling', *Biography* 33.3 (2010), pp. 471–97.

Wise, Amanda. ' "It's Just an Attitude that You Feel": Inter-Ethnic Habitus Before the Cronulla Riots' in Greg Noble (ed.) *Lines in the Sand: The Cronulla Riots, Multiculturalism and National Belonging* (Sydney: Federation Press, 2009), pp. 127–45.

Wolfe, Patrick. 'On Being Woken Up: The Dreamtime in Anthropology and in Australian Settler Culture', *Comparative Studies in Society and History* 33.2 (1991), pp. 197–224.

Wolk, Josh. 'Tomb Reader', *Entertainment Weekly* 614 (14 September 2001), pp. 57–9.

Yates, Frances A. *The Art of Memory* (Chicago: University of Chicago Press, 2001 [1966]).
Young, James E. 'The Counter-Monument: Memory Against Itself in Germany Today', *Critical Inquiry* 18.2 (1992), pp. 267–96.
Zizek, Slavoj. *Welcome to the Desert of the Real: Five Essays on September 11 and Related Dates* (London: Verso, 2002).
Żmijewski, Artur. *80064* (2004), Poland.

Index

Note: Page numbers in italics indicate figure.

A.#02 Avignon, 144
Aboriginality, 57, 60, 61, 64
  colonial power structure as, 69, 75
  performing, 65–7, 72, 74
Absent Synagogue, The, 29
aesthetics, 19, 44
  digital, 116
  perceptual processes and, 155, 156
  practical, 139
  remembrance of, 40
  theories of, 19
  transmission of affects, 109, 139
affective spill, 5, 6, 26, 27, 71, 99, 103, 105, 136–57
Ahmed, Sara, 3, 4, 5, 6, 17, 41, 55, 57, 58, 62, 90, 115, 132, 152
a-materiality, 26, 107–35
Apel, Dora, 17, 40
Apter, Emily, 120, 129
archive, 97, 137
  affects of, 136–42
  archival culture, 14, 97
  public sentiment of, 115
Assman, Jan, 13
atom bomb, *see* Hiroshima
audience, *see* spectator
Auschwitz–Birkenau, 1
  see also Auschwitz Death Camp
Auschwitz–Birkenau Memorial Museum, 107
Auschwitz Death Camp, 7, 29
Australia, 24, 56–78
avatar, 26, 117, 122, 124, 126, 128–32, 135
  see also Second Life

B.#03 Berlin, 144
Barnum, Phineas T., 65

Bartov, Omer, 40, 42
Bennett, Jill, 139, 151, 152, 155, 156, 157
Berlant, Lauren, 17, 18, 31
Best, Susan, 42, 43
biosocial, 12
Bishop, Clare, 2
body, the, 2, 3, 4, 8, 10, 11, 12, 18, 19, 20, 21, 31, 38, 42, 43, 48, 51, 55, 64, 90, 119, 129
  biological, 112, 134
  corporeal, 6, 10, 21, 116, 139, 140, 146, 151; inter–corporeal, 38, 39, 56–78; post-corporeal, 38, 39; syncope, 42–3
  embodiment, 5, 7, 9, 10, 15, 23, 31, 32, 33–7, 38, 62, 108, 109, 119, 126, 127, 129
  emotion and, 4, 20, 22
  iconography of suffering as, 149
  memory, 14, 21, 105, 118, 126
  social construction of, 62
  somatic marking, 5, 21, 126, 139
  supra-body, 10
  virtual, 134; *see also* avatar, Second Life
Boltanski, Christian, 23, 31, 41–7, 49
Bolter, Jay, 114
BR.#04 Bruxelles, 136, 140, 141, 146, 147, *148*, 149, 151, 157
Brennan, Theresa, 12, 22, 38, 42, 43, 44, 48
Butler, Judith, 6, 9, 112, 116
Butler, Kelly Jean, 58, 59

Callard, Felicity, 20
Camillo, Guillio, 142, 146
Caruth, Cathy, 15, 113, 142

Casey, Edward S., 20–1, 126
Casey, Maryrose, 74
Castellucci, Romeo, 137, 138, 144, 146, 149
catharsis, 10, 18, 33, 80, 87
chronotopic suspension, 45
Clarapède, Édouard, 139
Clark, Laurie Beth, 7
Clark, Marshall, 104
Clough Patricia Ticineto, 10, 12, 23, 118, 120
cognition
   affective memory and, 4
cognitive psychology, 20
Cohen, Michael, 61
collective
   audience as, 89
collective memory, 6, 7, 13–14, 99, 107
colonialism, 24, 57, 62, 66
   performance and, 74
   terra nullis and, 76
concentration camp memorials, *see* memorial
Connerton, Paul, 7, 8
contagion, affective, 48, 54, 71, 72, 154
   dynamics of, 64
Corporeal, *see* body, the
corroboree, 57, 74
   tourist, 66, 74, 75
Cowlishaw, Gillian, 77
Crossing Over with John Edward, 24–5, 79–106
Crownshaw, Richard, 13, 14, 31
Csordas, Thomas, 119
Cull, Laura, 156
Cunningham, Robert A., 65
Cvetkovich, Ann, 115
cyber commemoration, 107
cyber memorialisation, 107
Cyronic Chants, 147

Dachau Concentration Camp, 29, 31, 33, 34, 37, 39–41, 45

Dachau Concentration Camp Memorial Site, 23
Damasio, Antonio, 22, 119, 126
dance, 29, 31, 52, 53, 57
Dancing Auschwitz, 23, 28, 29, 31, 32, 53, 54, 55
Dapin, Mark, 38
Decreus, Freddy, 145
de-remediation of affective memory, 155–7
Dibbell, Julian, 108, 110, 111
digital history, 107
digital interfaces, 25
digital practice, 107
   see also cyber commemoration; cyber memorialisation
Dimoch, Wai Chee, 109
Djapugay people, 67, 69
Dovey, Jon, 95, 97
Dreaming, 24, 57, 59, 61, 71, 76
Dreamtime, *see* Dreaming

Edward, John, 79–106
   scepticism, 83–4
80064, 1, 2, 9
Eisenman, Peter, 23, 31, 49–52
Elder, Catriona, 61
Embodiment, *see* body, the
emotion, 3, 4, 5, 10, 11, 13, 19, 22, 38
empathy, 17, 24, 41, 150–2
episodic memory, 20
Erll, Astrid, 114
Escolme, Bridget, 138
ethics, 2, 15, 17, 18, 26, 30, 32, 39
   cultural memory of, 151
   digital, 110
   ethical citizenship, 59
   sexual, 110
   theatre as, 72
ethno-tourism, 24, 57, 67–70
   curatorial decision making and, 69
   models, 69
   target market, for, 68
ethnography, 35

feminist theory, 10
fire
   Indigenous ritual, in, 72–6
fireworks
   symbolism of, 74, 75
Fischer-Lichte, Erika, 138
Fisher, Jennifer, 19
Fison, Lorimer, 64
Fordyce, Ehren, 138
Frank, Anne, 107, 133
Freeman, Cathy, 61
Freud, Sigmund, 20

Garde-Hansen, Joanne, 114
Gaynor, Gloria, 29
Generational, *see* intergenerationality
gestures
   memory and, 29, 32, 36, 48, 52, 54, 55, 61, 62, 88, 108, 112, 117, 126, 144
ghosts, 25, 66–80
   demise of, 84–5
   ghost dogs, 82–6
   postmodern, 84–5, 86, 89
   see also supernatural; psychic
Gibbons, Scott, 146
Gibbs, Anna, 11, 48
Goodall, Jane, R., 66
Gordon, Avery, 81, 84, 85, 113
grief, 9, 24, 25, 79, 81, 96, 97, 104, 153
Grindstaff, Laura, 96
Grusin, Richard, 91, 92, 114
Guilio Cesare, 138
Guinness Book of Records, 24, 57

Halbwachs, Maurice, 13
Hamilton, Ann, 52
Hansen, Mark, 119, 120, 129, 134, 156
Hardt, Michael, 10
Harris, Geraldine, 37
Healy, Chris, 57
Hemmings, Clare, 44
Highmore, Ben, 19

Hiroshima, 115, 142
Hirsch, Marianne, 10, 15, 118, 128
Holocaust, 8, 17, 23, 24, 28–55, 107, 108, 132
   digital media and, 107
   post-Holocaust visual art, 10
   survivors, 6; tattoo, 1, 2, 6, 7, 8
hormonal processes, memory and, 12
Hrvatin, Emil, 48
Huyssen, Andreas, 14, 108

indexicality
   hyper indexical, 128, 130
   material indexical, 128
   post indexical, 51–2, 128, 142
Indigenous Australia, 24, 56–78
industrialisation, 20
intergenerationality, 10, 15, 16, 23, 31, 33, 123
I Will Survive, 29
IWitness, 107

Jameson, Frederic, 120, 142
Jewish identity, 40, 45

Kadishman, Menashe, 23, 31, 46, 48, 49, 51
Kansteiner, Wulf, 13, 14, 23
Kaplan, E. Ann, 115
Kelleher, Joe, 137, 141, 149
kinaesthetic, 118, 151
Kirby, Vicki, 6
Kirshenblatt-Gimblett, Barbara, 34
Korman, Jane, 28, 53
Kotthoff, Helga, 153
Kristallnacht, 108–10
   Second Life, *121, 123, 125, 130, 133*

LaCapra, Dominic, 41
Lacey, Kim, 44, 126
Landsberg, Alison, 16, 17, 18, 88, 118
Langton, Marcia, 75
Lazonga, Vyyn, 7

Lehmann, Hans-Thies, 143, 147
Lodz Ghetto, 29
loss, 6, 9, 13–19, 26, 27, 34, 35, 36, 44, 51, 52, 81, 102, 108
  Holocaust and, 29–30
  loss of loss, 120, 129, 141
  morality and, 31, 41, 83
  mourning and, 89, 104
  narratives of, 81
  poetics of, 134

Maisel Synagogue, The, 29
Malkin, Jeanette, 142, 143
Mandel, Naomi, 30, 32, 41
Marker, Chris, 16
Massumi, Brian, 10–11, 22, 42, 43, 44, 90, 118, 120, 151
memes of signification, 90
memorial, memorialisation, 14, 4, 33, 36, 52, 53, 104
  concentration camp memorial, 124
  politics of, 124
  remembrance, 50
  virtual, 134
Memorial to the Murdered Jews of Europe, 23, 31, 49, 50
Memory-Theater and Postmodern Drama, 142
Missing House, 23, 31, 41–6, *43*
Mitchell, Thomas, 74
Muecke, Stephen, 57, 78
multiculturalism, 58
Munch, Edvard, 56, 57
museum, 4, 14, 26, 29, 109, 128
  virtual, 117, 124, 129

Nancy, Jean-Luc, 140, 152–3
narrative, 88–9
national memory, 57, 60, 63
Nazism, 40, 45
Nietzsche, Friedrich, 90
Nora, Pierre, 14, 142, 107–8
nostalgia, 14, 50
Novati, Gabriella, 138

Olfactory, *see* smell
Olympic opening ceremonies, 60–1, 64
online communities, 110
otherness, 56–76
  nationhood and, 56

pain, pain affect, 5, 90, 91
  aesthetics of, 153
  representation of, 148
  see also trauma
Papoulias, Constantina, 20
Paranormal, *see* supernatural
Parsons, Michael, 66, 74, 75
Patraka, Vivian, 29, 30, 108, 131
Pellegrini, Ann, 115
performance theory, 35
pet heaven, 82–3
Phelan, Peggy, 35, 36
Phenomenology, *see* body, the
physiology, *see* body, the
Poignant, Roslyn, 65
politics, 24, 26, 27, 31, 33, 40, 101, 151
  cultural, 5
  feeling of, 33
  pain of, 5
  sentimental, 17, 18, 24, 31
  theatre as, 72
postmemory, 10, 15, 16, 33
postmodernity, 14, 80, 82, 120
  ethics and, 137
  politics and, 137
Potts, John, 84
Povinelli, Elizabeth, 58, 63, 64
practitioning, cultural memory of, 6, 14
premediation
  theory of, 91–3
Pruchnic, Jeff, 44, 126
psychic, 16, 18, 26, 24, 79–106
  residence, 18
psychomsomatic, 12, 19, 48
pyrotechnics, *see* fireworks
pyschodynamics
  affect of, 19, 48

queer theory, 10

Radgast Train Station, 29
rape, virtual, 110–11
Reading, Anna, 112–13, 116
reality television, 95–6, 98
   pseudo-documentary as, 98
reconciliation, 61
Redfield, Marc, 114, 115, 132
remediation, 79–106
   definition, 114
   digital media and, 114
   hyper–remediation, 110
   medium, 91
   memory affect of , 26, 30, 49, 81, 82, 108–9, 128, 148
   Second Life and, 108
   television and, 81, 92
repatriation, 66
Ribot, Théodule, 20, 21
Ridout, Nicholas, 18, 71, 72, 137, 141, 145, 147, 149
Rigney, Ann, 114
Rigney, Daryle, 61
Roach, Joseph, 35, 49
Rowe, Deborah, 60, 61

Sack, Daniel, 147
Sagir, Eli, 6, 8
Saltzman, Lisa, 10, 51, 127, 142
Sampson, Steven, L., 42
Sans Soleil, 16
Scarry, Elaine, 2, 150
scenario
   cultural repertoires as, 61, 62, 63
Schmitt, Canon, 116, 129
Schneider, Rebecca, 8, 15, 23, 32, 35, 36, 55, 138
scream, *see* sound
séance, 86
Second Life, 4, 25, 107–35
Second Life Kristallnacht, *see* Kristallnacht
Second Life museums, *see* museums, virtual
Sedgwick, Eve Kosofsky, 11, 22

self
   Cartesian, 12
   self-making, 3, 19, 20, 23
Seltzer, Mark, 115, 133
semiotics, 76, 87–94, 102, 124, 133
Senft, Theresa, 108, 111, 112, 116, 124, 131, 134
sensory, 22–7
   memory, forms of, 4, 13, 19, 29, 31, 32, 33, 34, 38, 39, 44, 45, 64, 108, 153
   politics of, 155
sensuality, 4
   tautology and, 144
sentiment, 11, 13, 24, 29, 31, 33, 38, 81, 89
September 11, 25, 82, 92, 93, 103, 104, 114
Seremetakis, C. Nadia, 21
Shalechet (Fallen Leaves), 23, 31, 46–9, *47*
Shoah, *see* Holocaust
Shoah Foundation, 107
   see also IWitness
Shouse, Eric, 11, 12
Silverman, Kaja, 16, 18, 118
smell, 3, 12, 38, 39, 54
Societas Rafaello Sanzio, 26, 138–46
Socio-cultural process
   memory as, 21–2
Solomon-Godeau, Abigail, 41, 45
Somatic, *see* body, the
Sontag, Susan, 148, 149
sound, 54, 89, 149
   screams, 147; *Shalechet* and, 48
sound art , 146
spatial memory
   configuration of, 143
spectator, 25, 26, 30, 31, 32, 34, 36, 44, 51, 59, 66, 71, 91, 96, 98, 108, 109, 137, 152–5
   archive as, 150
   collective, 81, 89
   engagement, 109
   inversion of, 138
   participant as, 81, 87, 91

relationship with subject, 3
subjectivity, 93
witness as, 115, 145, 146
spiritualism, 86
Stephenson, John, 60, 61
Stolen Generation, 77
Stoller, Paul, 35
storytelling, 62
supernatural, 85–6, 96
  politics of, 86
  technology and, 85
Sutton, John, 20

Tambo, 65, 66
Tarnawa, Jósef, 1, 7
tattoo, 2, 6–9, 27
  cross-generational, 7
  physical boundary as, 7
  re-tattoo, 9
  signifying trauma, 2, 6, 7
Taylor, Diana, 35, 49, 61
television, 4, 24, 79, 95
  time and, 92
  see also reality television
Temporal, *see* time
Terezin Ghetto Fortress, 29
The Scream, 46–7
theatre, 4, 136–
  accidents, theatrical, 71–2
  aesthetics, 18; radical, 137
  perception, transformation of, 155
  politics of, 72
  postdramatic, 143, 147
Theresienstadt Concentration Camp, 29
time, 64, 131
  duality of, 45
  pre and post, 80
  televisual time and space, 91
  travel, memory as, 20–1
Tindale, Norman, 68
Titchener, E.B., 20
Tjapukai Aboriginal Cultural Park, 24, 57–78
Tomkins, Silvan, 47, 48

touch, *see* sensory
tourist, 31, 34, 36, 37, 49
  behaviour, 37–8
  concentration camp, 24
  industries, 53
  touristscape, 29, 51
  see also ethno-tourism
Tragedia Endogonidia, 136, 138, 142, 144
trauma
  act of, 2, 7
  affect, 3
  catharsis of; *see* catharsis
  collective, 10, 15, 25, 110
  contagion, 81
  cultural spaces, 15
  culture, 141–2
  discourse, 132, 150
  inherited, 15, 16, 54
  latency of, 113, 115
  memorial, 36
  memory, 3
  national, 103
  politically constituted, 81
  public phenomenon as, 33
  remediation and, 113
  social, 8
  spectator and, 154
  television, 4
  tourism, 53, 118
  traumascape, 127
  traumatic affect, 23, 55, 91, 134, 139
  traumatic memory, 21, 95
  transmissison of, 10, 59
  virtual, 108, 111, 112–16, 134, 145
  wound, 2; culture and, 109, 115
  see also BR.#04 Bruxelles
Trubridge, Sam, 138
Tumarkin, Maria, 127

United States Holocaust Memorial Museum, 25, 29, 108, 123
Urban, Greg, 153
Urban, Richard, 117

Vibratorium, 141
victim, 37, 38, 39, 42, 131
　histories, 27
　perpetrator, 131
Viewer, *see* spectator
violence
　media technologies and, 132
　virtuality of, 121

Web, 2.0, 112
Weissman, Gary, 40
welle, 52

Whitlock, Gillian, 109
Wise, Amanda, 19
Wolfe, Patrick, 76
Wolk, Josh, 101
Wound, *see* trauma

Yates, Frances, 143
Young, James E., 51

Zizek, Slavoj, 48, 104, 105
Zmijewski, Artur, 1, 3, 5, 8, 9
Zürcher Theatre Spektakel, 136, 137

GPSR Compliance
The European Union's (EU) General Product Safety Regulation (GPSR) is a set of rules that requires consumer products to be safe and our obligations to ensure this.

If you have any concerns about our products, you can contact us on

ProductSafety@springernature.com

In case Publisher is established outside the EU, the EU authorized representative is:

Springer Nature Customer Service Center GmbH
Europaplatz 3
69115 Heidelberg, Germany

www.ingramcontent.com/pod-product-compliance
Lightning Source LLC
Chambersburg PA
CBHW061806110426
42873CB00042B/44